THE HEAT PIPE

Other titles in this series

TL/ME/1 HYDROSTATIC EXTRUSION
J M Alexander
B Lengyel

Mechanical Engineering Monographs available

ME/1 TRIBOLOGY
E D Hondros

ME/2 ELECTRON BEAM WELDING
M J Fletcher

ME/3 VACUUM BRAZING
M J Fletcher

ME/4 STRESS CORROSION FAILURE
P Greenfield

All published by Mills & Boon Limited

M & B Technical Library, TL/ME/2

General Editor: Gordon Cook, Ph D

The Heat Pipe

D Chisholm, B Sc, Ph D, A R C S T, C Eng, M I Mech E

Head of Applied Heat Transfer Division, National Engineering Laboratory

Mills & Boon Limited

London

First published in Great Britain 1971
by Mills and Boon Limited, 17–19 Foley Street,
London W1A 1DR

ISBN 0 263 51711 X

Made and printed in Great Britain
by Butler & Tanner Limited, Frome and London

CONTENTS

PREFACE

This monograph is intended to serve two functions: first, to answer questions such as "What is a heat pipe?" and "How does it work?"; second, to serve as an introduction to the technical problems involved in the design and manufacture of heat pipes.

Chapters 1, 6 and 8 together will, it is hoped, meet the first of these requirements. The remaining chapters are concerned with the more detailed technical problems involved in design and manufacture.

SI units have been used in their basic form, with a few exceptions where it was felt that the reader would find other units more acceptable. A conversion table is given in Appendix I. The exceptions relate first to the use of heat flux as kW/cm^2, rather than kW/m^2. It was felt that this would give the reader a better "feel" for the magnitudes of the rate of heat transfer in heat pipe situations, as the relevant flow cross-sections are in practice normally only a few square centimetres in area. For a similar reason, heat pipe diameters are quoted in centimetres. The unit of pressure has been taken as the bar or, where high vacuum is discussed, the torr; the closeness of the relation between 1 bar and atmospheric pressure makes it attractive. The Kelvin (K) temperature scale has been used, except where it was more convenient or relevant to use the Celsius scale (°C).

In those chapters where a large number of symbols are used, the nomenclature appears at the beginning of the chapter; elsewhere, symbols are defined where they are first used.

In the bibliography, the reference system consists of the first letter of the first author's name followed by a number corresponding to the reference's alphabetic placing. The first entry under S is S1, and so on. In the text, reference is made with no redundancy of information; thus if an article is referred to in the text by the author's name, and the bibliography contains only one entry under that name, no bibliography reference number is given. Also, it is superfluous to write, for example, as a reference Busse[(B6)]; the statement Busse[(6)] is without ambiguity.

The monograph is published by permission of the Director, National Engineering Laboratory, Department of Trade and Industry.

1. Introduction

The heat pipe is a device which transfers heat by boiling a fluid at one point and condensing it at another; the liquid is transferred back to the boiling area by capillary action through a wick structure. It is this use of the capillary action which is the unique characteristic of the heat pipe.

The thermosyphon in many ways resembles a modern heat pipe, except that the return of the liquid to the evaporator is due to gravity. Fig. 1.1 illustrates the basic layout of the thermosyphon. The liquid boils in the lower regions of the cylinder and condenses in the upper regions, the condensate returning by gravity to the boiler or evaporator. It is, in fact, a reflux condenser.

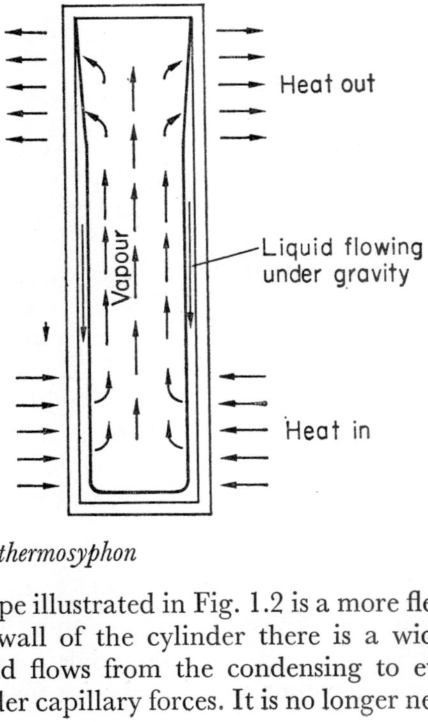

Fig. 1.1. A thermosyphon

The heat pipe illustrated in Fig. 1.2 is a more flexible unit. Along the wall of the cylinder there is a wick through which liquid flows from the condensing to evaporating sections under capillary forces. It is no longer necessary for the evaporator to be beneath the condenser, and operation of the device can be carried out satisfactorily with the major axis of the device in planes other than the vertical.

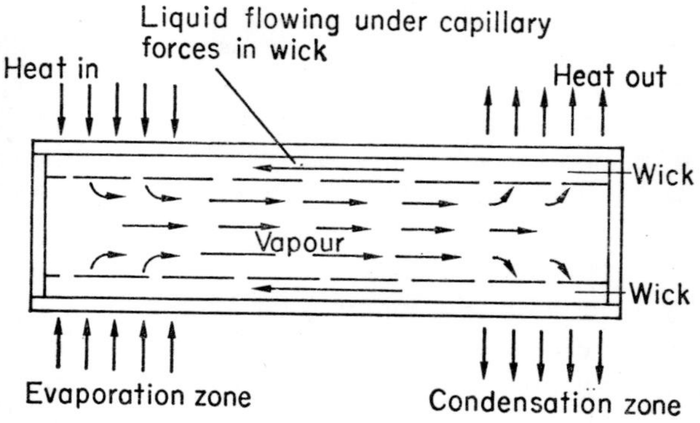

Fig. 1.2. A heat pipe

The effective thermal conductivity of the device can be as much as 500 times that of a solid copper rod of similar mass. Heat fluxes as high as 15 kW/cm^2 have been measured(B10) with lithium as the working fluid at temperatures of 1500°C.

The fluids and their pressures are determined by the operating temperatures of the device. Table 1.1 shows(F3) some properties of working fluids in heat pipes.

The first heat pipe patent, dated 1942, was issued to R. S. Gaugler (US Patent 2350348). The proposed application related to the cooling of the interior of an ice box. However, the pioneer work in this field was carried out by Grover(12) working at Los Alamos, with the object of using these devices in spacecraft; the absence of gravitational forces improves their performance in space, and much of the early work in this field was concerned with space applications.

In the simple type of heat pipe illustrated in Fig. 1.2 the wick is held tightly against the container wall. The original wick structures were materials such as woven cloth, fibreglass, porous metal and wire mesh; channels

Table 1.1. Fluids suitable for heat pipes (properties at atmospheric pressure)

FLUID	MELTING POINT K	BOILING POINT K	DENSITY* Kg/m^3	LATENT HEAT KJ/Kg	SURFACE TENSION mN/m (at melting point)
Ammonia	196	240	682	1370	41†
Water	273	373	1000	2256	76
Cesium	302	978	1794	612	76
Potassium	337	1033	819	2077	86
Sodium	371	1156	929	4210	190
Lithium	452	1590	509	19631	386
Lead	600	2010	10492	858	470
Indium	429	2360	6984	1963	559
Silver	1233	2485	9291	2326	920

*At melting point †At 244 K

were then used, which were later screened to eliminate or reduce entrainment of the liquid in the vapour. It was soon appreciated however that a composite wick with small pores at the liquid–vapour interface during operation, and large pores over the remainder of the wick, would give high capillary forces with reduced flow losses. The development of wicks along these lines has enabled heat pipe loadings to be increased(K4) by ten times.

The use of capillary pumping in two-phase systems has been considered in more complex geometries than the simple heat pipe configuration, Fig. 1.2. Laub and McGinness in 1961 proposed the use of a Freon two-phase circuit with a capillary pump for use in a gas bearing, as illustrated in Fig. 1.3. The Freon liquid is vapourized within

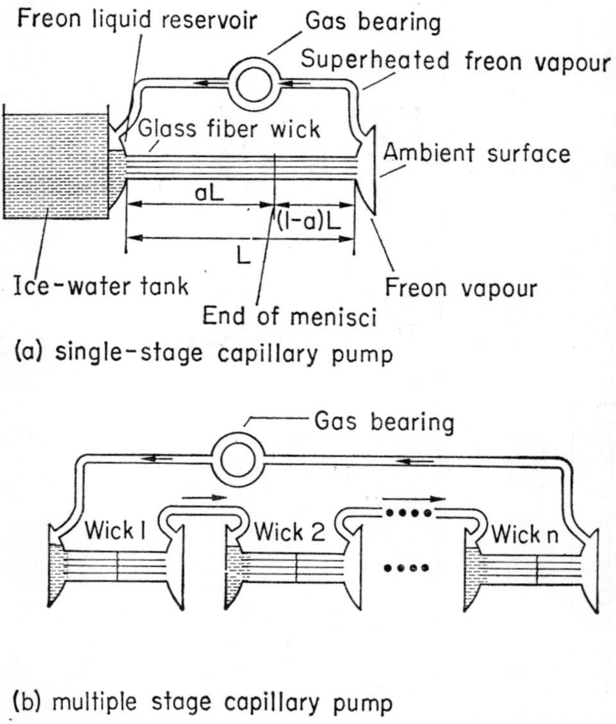

Fig. 1.3. Single-stage and multiple-stage capillary pumps. (Laub & McGinness)

the capillary structure; this is necessary otherwise the interface between the phases will not be within the fibre wick and the capillary pressure driving force will not be obtained. Similarly, between each stage in the multi-stage capillary pump also illustrated in Fig. 1.3, it is necessary to recondense the vapour.

The heat pipe normally contains fluid such that at the operating conditions there is a slight surplus of liquid to that required to saturate the wick. Fig. 1.4 shows diagrammatically the location of the interface between the phases

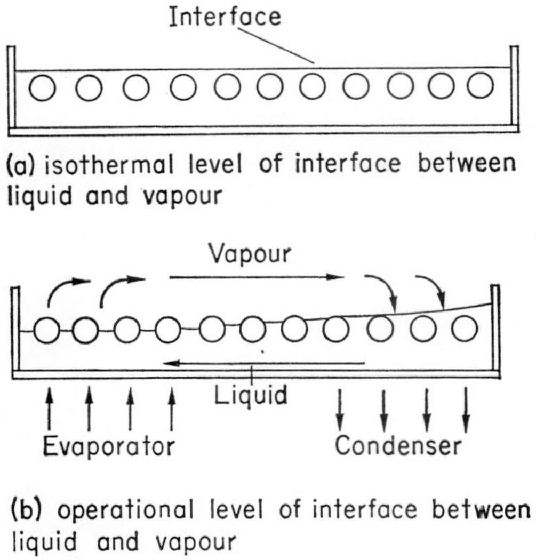

Fig. 1.4. Isothermal and operational levels of interface between vapour and liquid

for both isothermal and operational conditions. In the evaporating section, evaporation causes the radius of curvature at the miniscus to increase while condensation at the condenser results in the radius of curvature remaining infinite there; this difference in radius of curvatures provides the capillary driving force which circulates the working fluid. The pressure difference created is

$$\Delta p = 2\sigma\left(\frac{\cos \alpha_e}{R_e} - \frac{\cos \alpha_c}{R_c}\right) \tag{1.1}$$

where Δp is the pressure difference created by capillary forces
σ is the surface tension
α is the contact angle
and R is the radius of the capillary pore. The subscript e refers to properties in the evaporator and c to properties in the condenser.

This pressure difference is available to overcome resistance to circulation due to friction, changes of momentum, and gravity.

The axial rate of heat transfer is given by

$$Q = \dot{m}.L \qquad (1.2)$$

where Q is the rate of heat transferred axially
$\dot{m}$ is the rate of circulation of fluid round the heat pipe, and
L is the latent heat of evaporation.

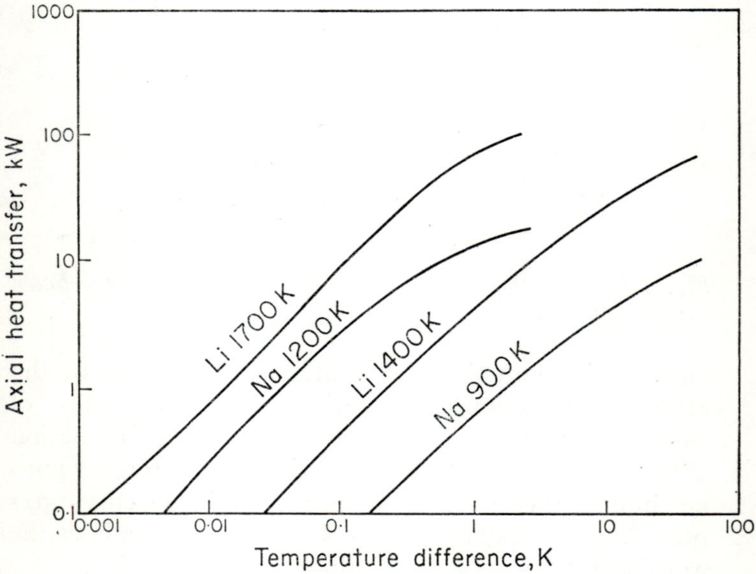

Fig. 1.5. Temperature difference over 0·3 m long heat pipe, 2·54 cm diameter as a function of the rate of heat transfer

This heat transfer rate is associated with a temperature difference within the vapour between the evaporator and condenser due to the vapour pressure difference between these points. Over the heat pipe there are in addition radial temperature differences in the pipe wall and due to the evaporation and condensation processes. Fig. 1.5 shows(F3) heat transfer rates and associated temperature differences over a heat pipe 2·54 cm diameter and 30 cm long with a woven mesh wick. Thus heat pipes can be designed to give very high rates of heat transfer for only a few degrees of temperature drop.

The output from a heat pipe can be limited by a number of considerations: the velocity of the vapour cannot exceed the sonic velocity; entrainment of liquid by the vapour must be avoided, or the evaporator will be starved of liquid; film boiling must be avoided due to the poor heat transfer coefficients obtained; and the final limitation is the rate of circulation of fluid obtainable. Fig. 1.6 illustrates(K5) diagrammatically how these four factors combine to give the operational envelope for a given design of heat pipe. At low vapour pressures the sonic velocity may

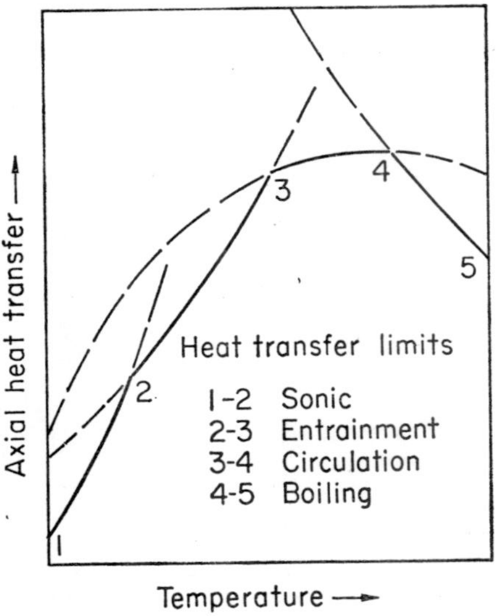

Fig. 1.6. Operational envelope: maximum heat transfer capacity as function of operating temperature

be the limiting factor as the gas density will be low; this is region 1–2 in Fig. 1.6. In region 2–3, the heat flow is limited by entrainment. In region 3–4, the restriction on output is due to limitations of the capillary action. In region 4–5 the axial heat flow is limited by the onset of film boiling.

The predicted curves in Fig. 1.5 were evaluated on the assumption that capillary action was the limiting criterion.

In addition to its ability to transfer high heat fluxes with small temperature gradients the heat pipe has the following advantages:

1. As a transformer it allows the heat to be absorbed as a high heat flux (i.e. over small area) and transferred from the heat pipe with a low heat flux (i.e. over a large area).
2. It serves as a temperature control device. By a number of methods it is possible to arrange the device to operate at an approximately constant temperature, to transfer heat only when required (the thermal switch), to allow heat flow in one direction only (the heat pipe diode), and to perform similar functions as illustrated in Fig. 1.7.

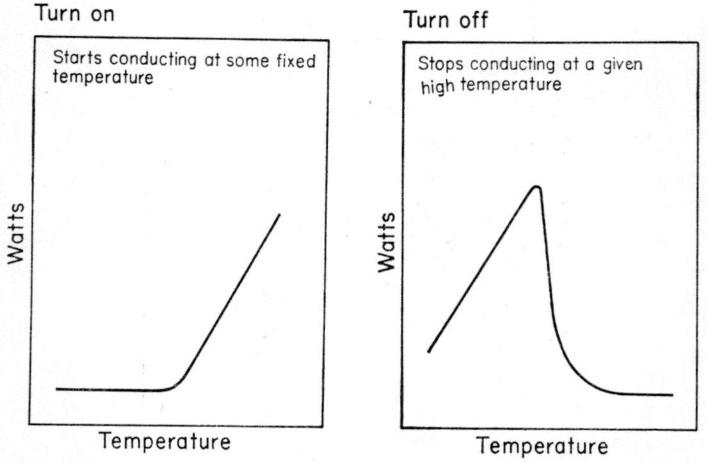

Fig. 1.7. Diagrammatic representation of characteristics of nine basic heat pipe functions. (*Original by Noren Products, Inc.*)

On and off

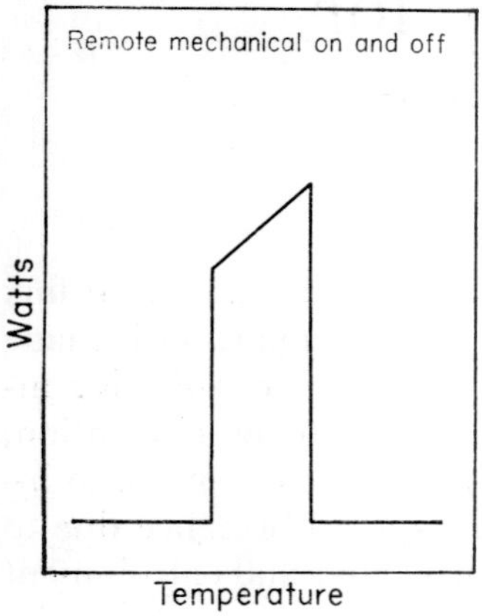

Isothermal

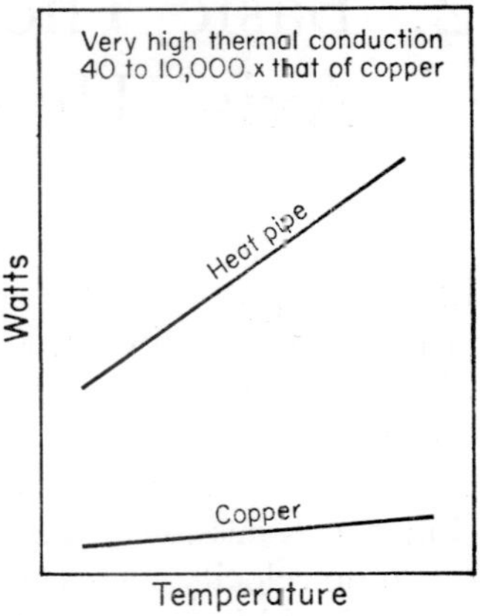

Maximum power

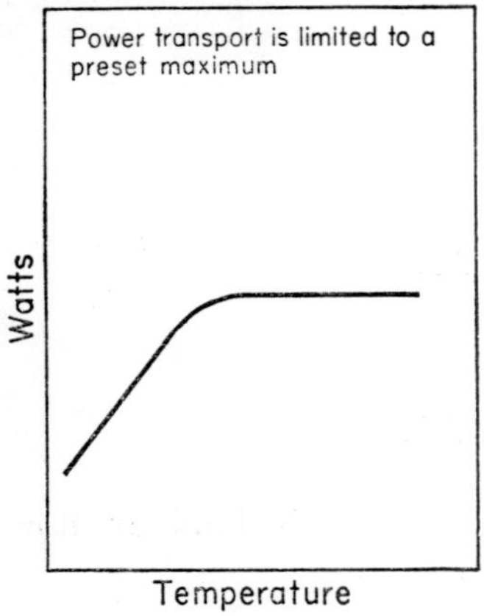

Two phase

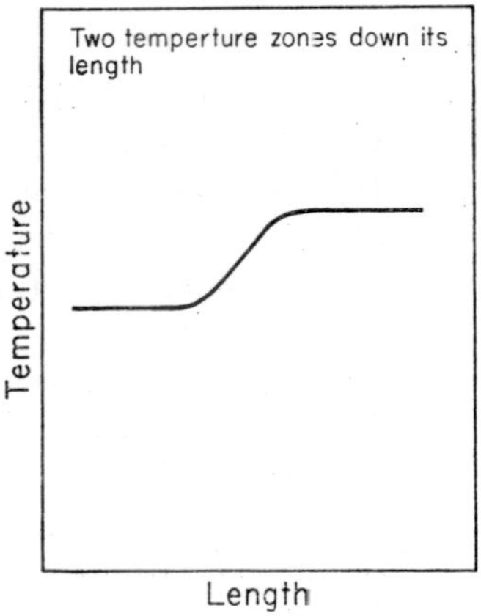

Heat Shield

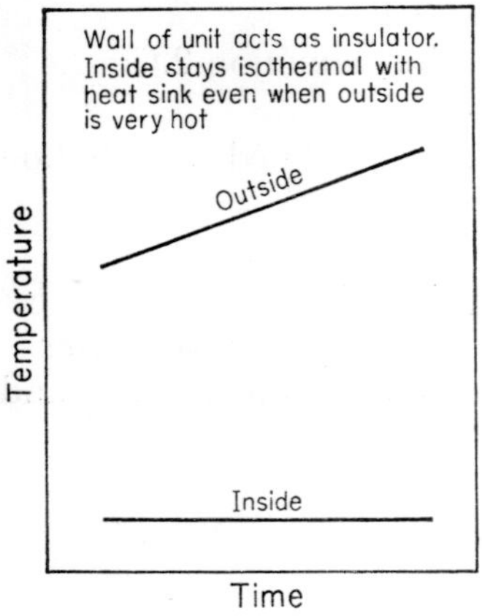

Thermal diode

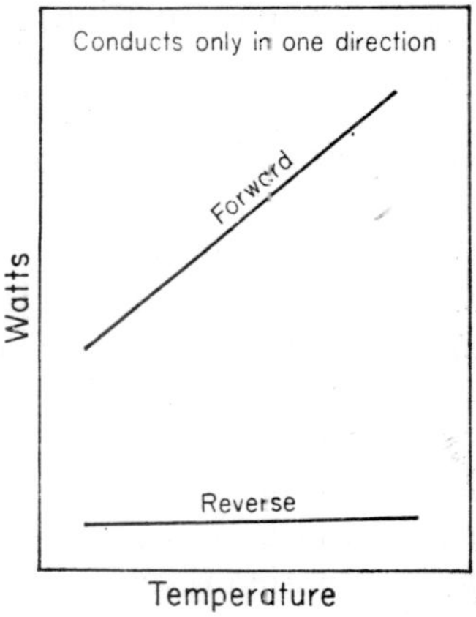

2. Basic Theory for Axial Heat Flux

In this chapter the means of predicting the axial heat flux in heat pipes is discussed, and some comparison is made with experiment. The greater part of the chapter is concerned with the prediction of the rate of circulation, assuming the fluid is incompressible; the chapter also includes a discussion of restrictions on performance due to the sonic velocity occurring in the vapour and entrainment of the liquid in the vapour.

Notation

A_{w}	Cross-section of wick
C	Coefficient in equation 2.23
D_{w}	Internal diameter of wick
g	Gravitational acceleration
K	Permeability (equation 2.5)
L	Latent heat
l	Length of heat pipe
l_{r}	Capillary rise (equation 2.12)
l_{c}	Length of condenser
l_{cl}	Length of condenser over which laminar flow occurs (equations 2.45 to 2.48)
l_{e}	Length of evaporator
l_{el}	Length of evaporator over which laminar flow occurs (equations 2.47 to 2.48)
l_{eff}	Effective length of heat pipe, equation 2.6
l_{S}	Length of shielded section
l_2	Effective length of heat pipe, defined by equation 2.38
M	Momentum flux
$\dot{m}$	Mass flow rate in shielded section
$\dot{m}'$	Local mass flow rate in evaporator or condenser
n	Ratio of specific heats at constant volume and pressure
p	Static pressure
p_{L}	Static pressure in liquid
p_{V}	Static pressure in vapour

Δp	Pressure difference
Δp_{c}	Maximum capillary pressure difference
$\Delta p'_{c}$	Capillary pressure difference at end of condenser
Δp_{fL}	Pressure difference in wick due to friction
Δp_{G}	Pressure difference due to gravity
Δp_{Lc}	Pressure difference in condenser for liquid
Δp_{Le}	Pressure difference in evaporator for liquid
Δp_{LS}	Pressure difference in shielded section for liquid
Δp_{Vc}	Pressure difference in condenser for vapour
Δp_{Vct}	Pressure difference in condenser in vapour over length where vapour flows turbulently
Δp_{Ve}	Pressure difference in evaporator in vapour
Δp_{VS}	Pressure difference in shielded section in vapour
Q	Axial rate of heat transfer
q	Axial heat flux
R	Radius of capillary interface
R_{e}	Radius of capillary interface in evaporator
Re	Reynolds number in axial vapour flow (equation 2.15)
Rr	Reynolds number in radial vapour flow (equation 2.20)
r_{v}	Internal radius of wick
U	Vapour velocity in axial direction
U_{s}	Vapour velocity in shielded section
U_{ch}	Sonic velocity of vapour
V	Vapour velocity towards interface
We	Weber number (equation 2.52)
Z	Characteristic dimension of liquid–vapour interface
α	Contact angle
θ	Inclination of heat pipe to horizontal: between zero and 180° from upward flow of liquid
λ	Friction factor
λ_{m}	Friction factor with mass transfer
μ_{L}	Absolute viscosity of liquid
μ_{V}	Absolute viscosity of vapour
ρ_{L}	Density of liquid
ρ_{V}	Density of vapour
σ	Surface tension

Subscripts

c	Condenser
e	Evaporator

f	Friction
G	Gravity
L	Liquid
S	Shielded
t	Turbulent
V	Vapour

2.1 THE CAPILLARY PRESSURE DIFFERENCE

At all points along the length of the heat pipe at the interface between the vapour and the liquid the difference in the static pressures of the phases is balanced by the local capillary pressure difference. The equilibrium condition is expressed

$$p_V - p_L = \frac{2\sigma . \cos\alpha}{R} \tag{2.1}$$

The interface adjusts its location within the wick pore structure such that there is an equilibrium condition; the interface finds the pore radius which satisfies equation 2.1.

Fig. 2.1 illustrates the variation of pressure and mass flow rate along the heat pipe and also the associated interfacial distribution. Maximum circulation will occur when there is no pressure difference between the phases at one point (in the case illustrated at the end of the condenser), while at the end of the evaporator the liquid interface is at the plane where the pore radius is smallest. Where there is no pressure difference between the phases at the end of the condenser the capillary pressure difference reduces to

$$\Delta p_c = \frac{2\sigma . \cos\alpha}{R_e} \tag{2.2}$$

The radius R_c is now infinite (see equation 1.1).

This capillary pressure difference in steady-state operation balances the friction, momentum and gravitational pressure differences in the phases. This can be expressed simply as

$$-\Delta p_c = \Delta p_V + \Delta p_L \tag{2.3}$$

Δp indicates a pressure rise in the direction of the flow of the phase (over the total length of the heat pipe unless otherwise indicated).

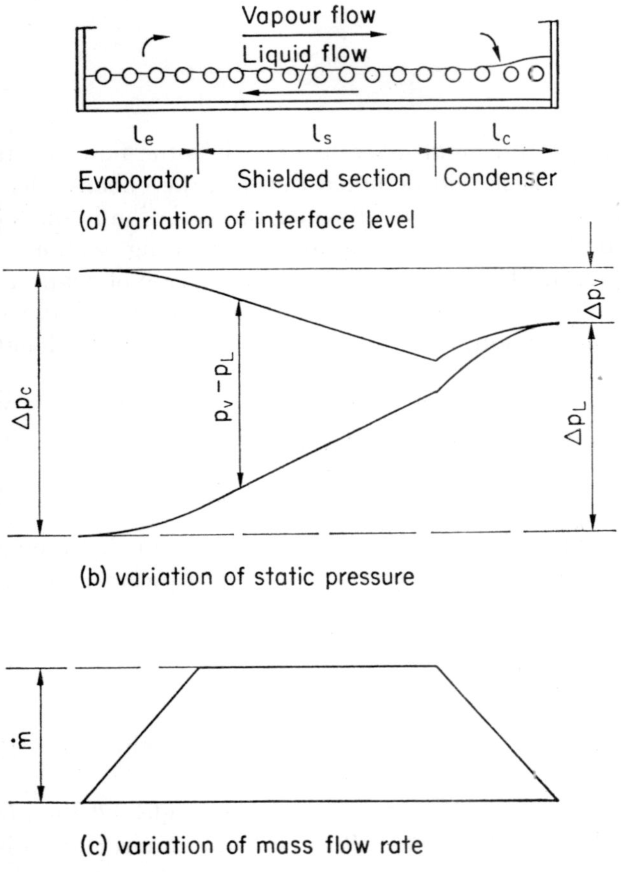

Fig. 2.1. Variation of interface level, static pressure and mass flow rate along length of heat pipe

2.2 PRESSURE DIFFERENCE IN THE LIQUID PHASE

The static pressure difference in the liquid can be expressed

$$\Delta p_L = \Delta p_{fL} + \Delta p_G \tag{2.4}$$

The pressure difference due to momentum changes can be neglected in the case of liquids. As discussed in more detail in Chapter 5, the pressure difference due to friction during laminar flow in a wick of porous material can be expressed as

$$\Delta p_{\mathrm{fL}} = -\frac{\mu_{\mathrm{L}}.\dot{m}.l_{\mathrm{eff}}}{\rho_{\mathrm{L}}.K.A} \tag{2.5}$$

Owing to the small pore size in wicks the flow is usually laminar. It should be noted that l_{eff}, the effective length, is used here. This is to allow for the variation in mass flow in both the evaporating and condensing sections (see Fig. 2.1). On the assumption that the rate of evaporation is uniform in both the evaporating section and the condensing section, it can be shown that the effective length is

$$l_{\mathrm{eff}} = l_{\mathrm{S}} + (l_{\mathrm{e}} + l_{\mathrm{c}})/2 \tag{2.6}$$

The pressure difference due to gravity is

$$\Delta p_{\mathrm{G}} = -g.\rho_{\mathrm{L}}.l.\sin\theta \tag{2.7}$$

The angle of the tube to the horizontal θ is between zero and 180° when the liquid is flowing upwards, i.e. when the evaporator is above the condenser.

From equations 2.4, 2.5 and 2.7

$$\Delta p_{\mathrm{L}} = -\frac{\mu_{\mathrm{L}}.\dot{m}.l_{\mathrm{eff}}}{\rho_{\mathrm{L}}.K.A} - g.\rho_{\mathrm{L}}.l.\sin\theta \tag{2.8}$$

If the pressure drop in the vapour is neglected then from equations 2.2, 2.3 and 2.8 the maximum circulation is

$$\dot{m} = \left(\frac{\sigma.\cos\alpha.\rho_{\mathrm{L}}}{\mu_{\mathrm{L}}}\right)\left(\frac{K.A}{l_{\mathrm{eff}}}\right)\left(\frac{2}{R_{\mathrm{e}}} - \frac{g.\rho_{\mathrm{L}}.l.\sin\theta}{\sigma.\cos\alpha}\right) \tag{2.9}$$

For this equation to give the maximum circulation rate, R_{e} must be the minimum effective pore radius in the wick. From equations 1.2 and 2.9 the rate of heat transfer along the heat pipe is

$$Q = \left(\frac{\sigma.\cos\alpha.\rho_{\mathrm{L}}.L}{\mu_{\mathrm{L}}}\right)\left(\frac{K.A_{\mathrm{w}}}{l_{\mathrm{eff}}}\right)\left(\frac{2}{R_{\mathrm{e}}} - \frac{g.\rho_{\mathrm{L}}.l.\sin\theta}{\sigma.\sin\alpha}\right) \tag{2.10}$$

In the absence of gravity, and making the usual assumption that there is perfect wetting ($\cos\alpha = 1$), this reduces to

$$Q = 2\left(\frac{\sigma . \rho_L . L}{\mu_L}\right)\left(\frac{K . A}{R_e . l_{eff}}\right) \tag{2.11}$$

It should be noted that the first group on the right-hand side is a characteristic of the fluid only, while the second is a characteristic of the wick. This is relevant to the selection of working fluids and wicks as discussed in Chapters 4 and 5.

In practice, the capillary pressure difference is obtained frequently by measuring the so-called "capillary rise"

$$l_r = \Delta p_c/(\rho_L . g) \tag{2.12}$$

as discussed in Chapter 5.

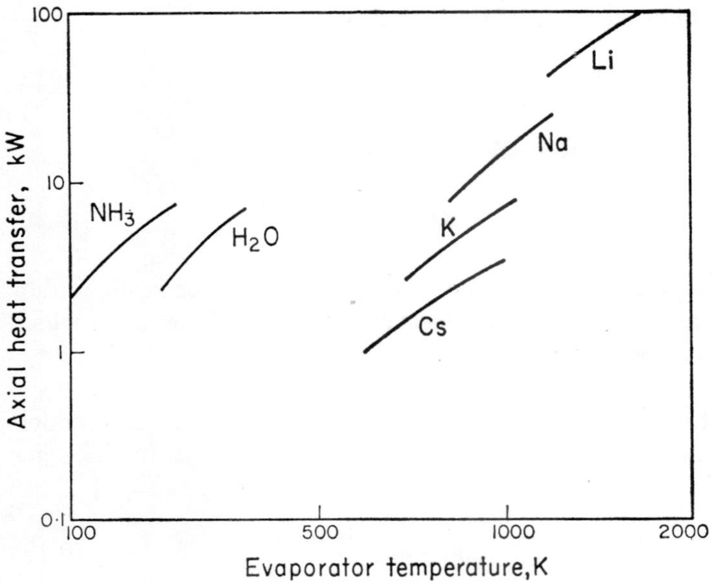

Fig. 2.2. Maximum heat transfer rate along 2.54 cm diameter heat pipe using equation 2.13

In terms of the capillary rise, equation 2.10 can be written

$$Q = \frac{g.\rho_L^2.K.A.L}{\mu_L.l_{eff}}(l_r - l \sin \theta) \tag{2.13}$$

and equation 2.11

$$Q = \frac{g.\rho_L^2.L.K.A.l_r}{\mu_L.l_{eff}} \tag{2.14}$$

Fig. 2.2 presents values(F3) of the maximum heat transfer capacity for a tubular heat pipe 2.54 cm in diameter and 30 cm long with a woven mesh wick obtained using equation 2.14.

2.3 PRESSURE DIFFERENCE IN THE VAPOUR PHASE

For the case where the vapour flows laminarly along the heat pipe in the shielded section (see Fig. 2.1) the pressure difference can be evaluated using

$$\Delta p_{VS} = -\frac{64}{\text{Re}}.\frac{U_S^2.\rho_V.l_S}{2\ D_V} \tag{2.15}$$

where the Reynolds number

$$\text{Re} = \frac{U_S.D_V.\rho_V}{\mu_V} \tag{2.16}$$

Equation 2.15 is the usual formula(M2) for laminar flow through pipes; it neglects a number of phenomena which will be discussed later.

In the literature on heat pipes, the commonest method(B4, C14) of expressing the vapour pressure difference is

$$\Delta p_{VS} = -\frac{8.\mu_V.\dot{m}.l_S}{\pi.\rho_V.r_V^4} \tag{2.17}$$

which is readily obtained by combining equations 2.15 and 2.16 with the continuity equation

$$U_S = \frac{\dot{m}}{\pi . r_V^2 . \rho_V} \tag{2.18}$$

In considering the effects of condensation and evaporation on the vapour flow, a further Reynolds number based on the radial flow of the vapour is defined

$$\text{Rr} = \frac{V . r_V . \rho_V}{\mu_V} \tag{2.19}$$

There is no significance in the characteristic length being the diameter in equation 2.16 and the radius in equation 2.19; this has become normal practice.

The radial velocity is related to the variation in axial mass flow rate of vapour as follows

$$V = \frac{1}{2\pi . r_V . \rho_V} . \frac{d\dot{m}'}{dl} \tag{2.20}$$

where the positive direction of l is measured in the direction of flow of the phase considered.

From equations 2.19 and 2.20

$$\text{Rr} = \frac{1}{2\pi\, \mu_V} . \frac{d\dot{m}'}{dl} \tag{2.21}$$

For $|\text{Rr}| \ll 1$ viscous effects dominate and the axial velocity profile is close to the usual parabolic shape for laminar flow. In this case the pressure gradient has been shown by Yuan and Finkelstein to be given approximately by

$$\frac{dp_V}{dl} = -\frac{8 . \mu_V . \dot{m}'}{\pi . \rho_V . r_V^4}\left(1 + \frac{3}{4}\text{Rr} - \frac{11}{270} . \text{Rr}^2\right) \tag{2.22}$$

Among other approximations, this equation neglects the effects of variations in Rr in the axial direction.

For $\text{Rr} \gg 1$ momentum forces dominate and the pressure gradient is given by

$$\frac{dp_V}{dl} = -\frac{c . \dot{m}'}{4 . \rho_V . r_V^4} . \frac{d\dot{m}'}{dl} \tag{2.23}$$

The coefficient C according to Cotter[14] is for evaporation unity, and for condensation $4/\pi^2$.

The case $|\mathrm{Rr}| \gg 1$, which is normally encountered in practice, is now developed in further detail. In Fig. 2.1 the mass flow variation illustrated is that of uniform rates of mass transfer to the wall in each of the sections in which heat transfer occurs; this corresponds to uniform rates of heat transfer. Therefore in the evaporator

$$\frac{d\dot{m}'}{dl} = \frac{\dot{m}}{l_e} \tag{2.24}$$

and in the condenser

$$\frac{d\dot{m}'}{dl} = -\frac{\dot{m}}{l_c} \tag{2.25}$$

Thus from equations 2.23 and 2.24 with Cotter's value[14] for the coefficient C the pressure drop over the evaporator in the vapour phase is

$$\Delta p_{Ve} = -\frac{\dot{m}^2}{8.\rho_V.r_V{}^4} \tag{2.26}$$

and in the condensing section from equations 2.23 and 2.25

$$\Delta p_{Vc} = \frac{\dot{m}^2}{2\pi^2.\rho_V.r_V{}^4} \tag{2.27}$$

Fig. 2.3 therefore represents diagrammatically the variation in vapour pressure in this case. The positive sign indicates that pressure recovery occurs in the condenser in the direction of vapour flow.

The resultant pressure difference, assuming there is no shielded section, is from equations 2.26 and 2.27

$$\Delta p_V = -\left(\frac{1}{4} - \frac{1}{\pi^2}\right)\frac{\dot{m}^2}{2.\rho_V.r_V{}^4}$$

$$= -0{\cdot}074\,\frac{\dot{m}^2}{\rho_V.r_V{}^4} \tag{2.28}$$

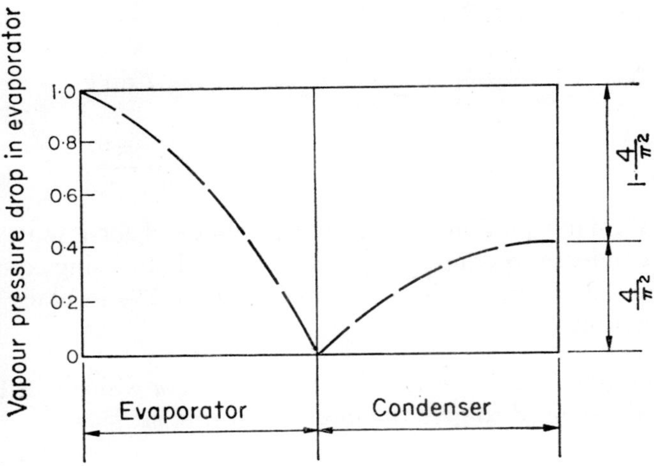

Fig. 2.3. Axial variation of vapour pressure. (*Cotter*, $R_r \gg 1$)

It is perhaps useful to transform this into some other forms which may be more familiar to the reader. It can readily be shown that in terms of the kinetic energy at the exit from the evaporator

$$\Delta p_V = -0{\cdot}149 \,.\, \pi^2 \,.\, \frac{U_S^2 \,.\, \rho_V}{2}$$

$$= -1{\cdot}47 \,.\, \frac{U_S^2 \,.\, \rho_V}{2} \qquad (2.29)$$

In terms of the momentum flux (rate of transfer of momentum per unit area) equation 2.28 can be written

$$\Delta p_V = -\frac{1{\cdot}47}{2} \,.\, U_S^2 \,.\, \rho_V$$

$$= -0{\cdot}735 \,.\, M \qquad (2.30)$$

If there is a shielded section then from equations 2.17 and 2.28

$$\Delta p_V = -\frac{8\mu_V \,.\, \dot{m} \,.\, l_S}{\pi \,.\, \rho_V \,.\, r_V^4} - 0{\cdot}074 \,.\, \frac{\dot{m}^2}{\rho_V \,.\, r_V^4} \qquad (2.31)$$

The pressure gradient due to gravity is usually small, and it has been neglected in obtaining this equation.

From equations 2.3, 2.8 and 2.31

$$\Delta p_c = \left(\frac{\mu_L . l_{eff}}{\rho_L . K . A} + \frac{8 . \mu_V . l_S}{\pi . \rho_V . r_V^4}\right)\dot{m} + g . \rho_L . l \sin\theta + 0{\cdot}075 \frac{\dot{m}^2}{\rho_V r_V{}^4} \tag{2.32}$$

A similar equation can be readily obtained for low rates of condensation and evaporation ($Rr \ll 1$) by using equation 2.22 as the basis of the prediction of Δp_V in place of equation 2.23.

Equation 2.32 is a quadratic equation in $\dot{m}$ which can be readily solved to give the circulation rate $\dot{m}$.

The equations discussed above are the most commonly used in predicting heat pipe performance (Dzakowic, Cotter,(14) Ernst). Busse(7) has shown that for values of Rr between -1 and ∞ the total vapour side pressure drop for the heat pipe is given by the normal equations for friction pressure drop in laminar flow, the net pressure drop due to momentum change being zero, the net effect of mass transfer on the wall friction also (surprisingly) being zero. That is

$$\Delta p_V = -\frac{8 . \mu_V . \dot{m}}{\pi . \rho_V . r_V^4} l_{eff} \tag{2.33}$$

where l_{eff} the effective length is given by equation 2.6.

Using equations 2.3, 2.8 and 2.33 the circulation is given by

$$\dot{m} = \frac{\Delta p_c - g . \rho_L . l . \sin\theta}{l_{eff}\left[\dfrac{\mu_L}{\rho_L . K . A} + \dfrac{8\mu_V}{\pi . \rho_V . r_V{}^4}\right]} \tag{2.34}$$

2.4 CIRCULATION WHEN $\Delta p_{Vc} > -\Delta p_{Lc}$

Fig. 2.1 has been constructed for the case where

$$\Delta p_{Lc} > -\Delta p_{Vc}$$

i.e. where the pressure recovery in the vapour flow direction in the vapour phase is less than the pressure drop in the liquid flow direction in the condenser.

If $\Delta p_{Vc} > -\Delta p_{Lc}$ and the point of zero curvature of the interface occurred at the end of the condenser, then the pressure distribution would be as shown in Fig. 2.4a. At vapour entrance to the condensing section $p_L > p_V$, hence at that point it would be necessary for a convex interface

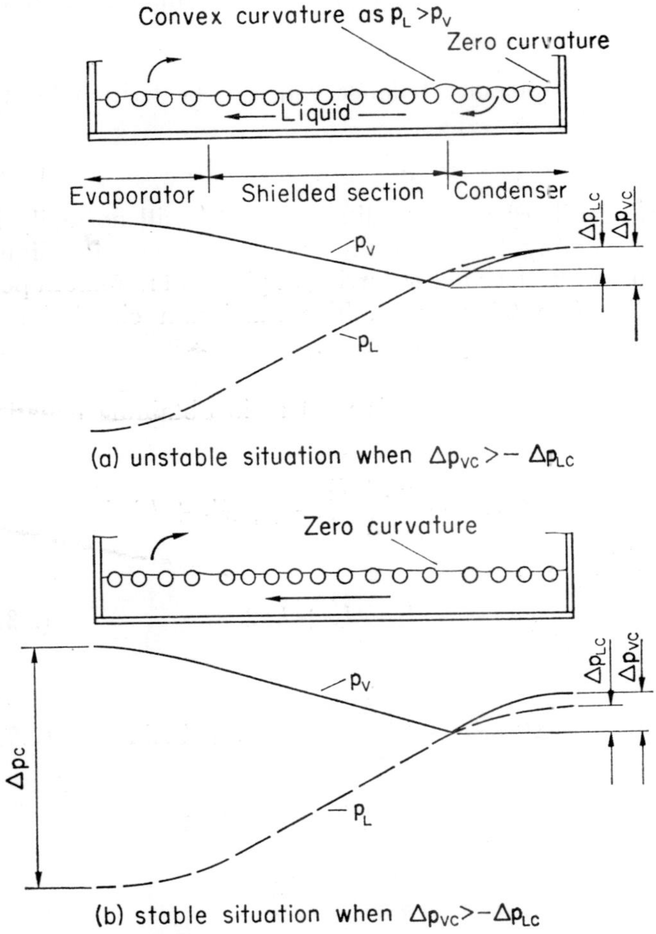

Fig. 2.4. Variation of static pressure and interface level when $\Delta p_{Vc} > -\Delta p_{Lc}$

to occur as shown in Fig. 2.4a. This will *not* occur under normal wetting conditions of the surface, and the equilibrium distribution will be(B5) as shown in Fig. 2.4b.

With this pressure distribution, circulation theory takes as its starting point the pressure balance

$$-\Delta p_c = \Delta p_{Ve} + \Delta p_{VS} + \Delta p_{LS} + \Delta p_{Le} \quad (2.35)$$

rather than the pressure balance represented by equation 2.3 which also includes the pressure differences in the condenser.

A similar pressure balance can also be written for the condenser

$$-\Delta p_c' = \Delta p_{Vc} + \Delta p_{Lc} \quad (2.36)$$

The capillary pressure difference $\Delta p_c'$ will normally be considerably less than the wick can exert, the liquid vapour interface not occurring at planes of minimum pore radius (see Chapter 5). The maximum circulation is determined by equation 2.35.

If a similar procedure is used as in obtaining equation 2.8 then

$$\Delta p_{LS} + \Delta p_{Le} = -\frac{\mu_L . \dot{m} . l_2}{\rho_L . K . A} - g . \rho_L (l_e + l_S) \sin\theta \quad (2.37)$$

where

$$l_2 = l_S + l_e/2 \quad (2.38)$$

and

$$\Delta p_{Lc} = -\frac{\mu_L . \dot{m} . l_c}{\rho_L . K . A . 2} - g . \rho_L . l_c . \sin\theta \quad (2.39)$$

From equations 2.17, 2.26, 2.35 and 2.37

$$\Delta p_c = \left(\frac{\mu_L . l_2}{\rho_L . K . A} + \frac{8 \mu_V . l_S}{\pi . \rho_V . r_V^4}\right)\dot{m} + g . \rho_L (l_e + l_S) \sin\theta + \frac{\dot{m}^2}{8 . \rho_V . r_V^4} \quad (2.40)$$

Thus if $\Delta p_{Vc} > -\Delta p_{Lc}$, that is from equations 2.27 and 2.39 if

$$\frac{\dot{m}^2}{2\pi^2 . \rho_V . r_V^4} > \frac{\mu_L . \dot{m} . l_c}{\rho_L . K . A . 2} + g . \rho_L . l_c . \sin\theta$$

then equation 2.40 is used to predict circulation rather than equation 2.32.

More advanced theoretical methods for predicting the vapour phase pressure gradient are presented by Busse.[6]

2.5 TURBULENT FLOW OF THE VAPOUR

With turbulent flow of the vapour which occurs approximately when Re > 2000, the pressure difference equation equivalent to equation 2.15 is

$$\Delta p_{VS} = -\frac{\lambda . U_S^2 . \rho_V . l_S}{2D_V} \tag{2.41}$$

where

$$\lambda = \frac{0{\cdot}316}{\mathrm{Re}^{0\cdot 25}} \tag{2.42}$$

With mass transfer the friction factor is modified[W3] to

$$\lambda_m = \frac{0{\cdot}316}{\mathrm{Re}^{0\cdot 25}} \pm 1{\cdot}4 \frac{V}{U} \tag{2.43}$$

This formula has been developed for the case of mass transfer to the wall, but it will also be assumed applicable, in the absence of evidence to the contrary, to mass transfer from the wall; the negative sign is applicable in this latter case. Silver and Wallis,[S10, W2] and Kinney and Sparrow have examined this problem in detail.

Neglecting effects associated with the variation in velocity across the flow section, the momentum flux between the evaporating and condensing sections is

$$M = U_S^2 . \rho_V$$

$$= \frac{\dot{m}^2}{\pi^2 . \rho_V . r_V^4} \quad (2.44)$$

It can be shown that the pressure difference from condenser inlet to the point of the onset of laminar flow in the condensing section is

$$\Delta p_{Vct} = -\frac{\lambda . \dot{m}^2}{4\pi^2 \rho_V . r_V^5} \left[\frac{l_c}{2{\cdot}75}\left\{1 - \left(\frac{l_{c1}}{l_c}\right)^{2{\cdot}75}\right\}\right] + \frac{0{\cdot}912 . \dot{m}^2}{\pi^2 . \rho_V . r_V^4}\left[1 - \left(\frac{l_{c1}}{l_c}\right)^2\right] \quad (2.45)$$

where

$$\frac{l_{c1}}{l_c} = \frac{2000}{\text{Re}} \quad (2.46)$$

The friction factor is evaluated at inlet to the condensing section. Equation 2.46 assumes that mass transfer does not influence the transitional Reynolds number. The final term in equation 2.45 combines the change in momentum flux over the turbulent length and the modification to the friction factor (equation 2.43). Over the laminar length the pressure drop can be evaluated as discussed earlier. The above procedure can also be used in the evaporating section.

Where the pressure difference over the entire length of heat pipe is required, then—assuming that the various effects of mass transfer cancel out—it can be shown that

$$\Delta p_V = -\frac{\lambda . \dot{m}^2}{4\pi^2 \rho_V r_V^5}\left[\frac{l_c}{2{\cdot}75}\left\{1 - \left(\frac{l_{c1}}{l_c}\right)^{2{\cdot}75}\right\} + \frac{0{\cdot}016 . l_{c1}}{\lambda}\left(\frac{l_{c1}}{l_c}\right)^2 + l_S + \frac{l_e}{2{\cdot}75}\left\{1 - \left(\frac{l_{e1}}{l_e}\right)^{2{\cdot}75}\right\} + \frac{0{\cdot}016 . l_{e1}}{\lambda}\left(\frac{l_{e1}}{l_e}\right)^2\right] \quad (2.47)$$

where

$$\frac{l_{e1}}{l_e} = \frac{l_{c1}}{l_c} = \frac{2000}{\text{Re}} \quad (2.48)$$

and λ is the friction factor in the shielded section.

Here, and in the previous sections, the equations for the pressure difference in the vapour neglect the effect of the interface being liquid rather than a rigid surface, and also neglect effects of the two-phases flowing in counterflow. Further information related to this can be obtained from the article by Bähr, Burck and Hufschmidt.

2.6 CHOKING

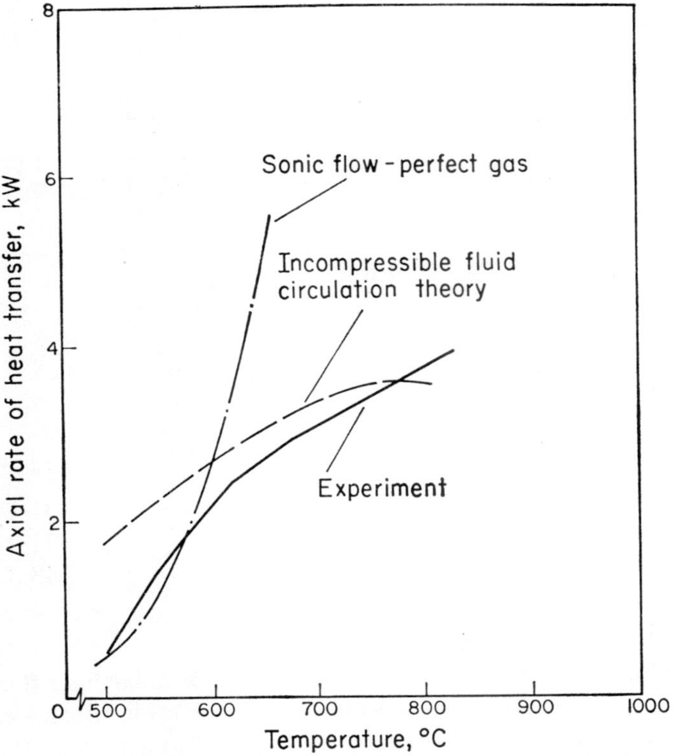

Fig. 2.5. Operational envelope when limited by choking and capillary pumping capability: comparison of theory and experiment. (Levy)

"Choking" refers to the condition when flow is restricted due to the sonic velocity occurring at some point in the flow path. This can happen at entry to the condenser in a heat pipe, in which case the circulation rate cannot exceed

$$\dot{m} = \pi . r_V{}^2 . U_{ch} . \rho_V \tag{2.49}$$

where U_{ch} is the vapour sonic velocity. The maximum axial rate of heat transfer is then

$$Q = \pi . r_V{}^2 . U_{ch} . \rho_V . L \tag{2.50}$$

For the case of a perfect gas

$$U_{ch} = \sqrt{\frac{n . p_V}{\rho_V}} \tag{2.51}$$

where n is the ratio of the specific heats.

Table 2.1: Sonic limitations of heat-pipe working fluids

EVAPORATOR EXIT	HEAT-FLUX LIMITS kW/cm²			
TEMPERATURE	Cs	K	Na	Li
400	1·0	0·5	—	—
500	4·6	2·9	0·6	—
600	14·9	12·1	3·5	—
700	37·3	36·6	13·2	—
800	—	—	38·9	1·0
900	—	—	94·2	3·9
1000	—	—	—	12·0
1100	—	—	—	31·1
1200	—	—	—	71·0
1300	—	—	—	143·8

Table 2.1 shows[(K5)] the limiting values of axial heat flux set by the occurrence of sonic flow. The table illustrates that the sonic or choking condition is of importance at the lower temperature limit of a fluid's range of operation, and therefore tends to be a factor of importance at start-up. Fig. 2.5 shows predicted axial heat fluxes for a sodium heat pipe compared with experimental values; the experimental values conform fairly closely to the envelope formed by the choking and circulation limiting conditions.

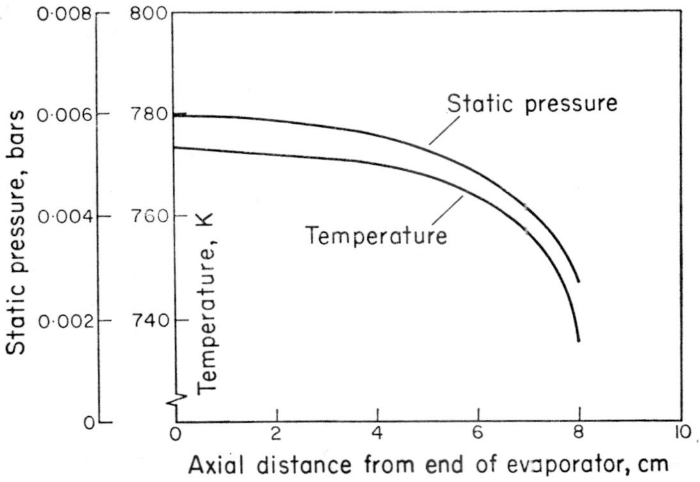

Fig. 2.6. Variation of temperature and pressure under choking conditions

The most difficult part[L6] of calculating the choking condition is in estimating the pressure at inlet to the condenser, as at the conditions under which choking occurs there can be a significant axial variation of temperature and pressure, as shown in Fig. 2.6.

2.7 ENTRAINMENT

As the vapour and liquid are in counterflow within the heat pipe, there is a possibility that the vapour will pick up or entrain liquid from the surface of the wick. Kemme[3] measured the axial heat flux of a sodium heat pipe with a composite wick consisting of 0·75 × 0·75 mm channels in the wall covered with a screen of 46 μ wires spaced 10 μ apart. The upper limit of performance was found to be set by the on-set of entrainment, which was detected by the sound of liquid droplets striking the end of the condenser section.

The criterion for the onset of entrainment[C7, K3] is that the Weber number should be greater than unity. The Weber number is defined as

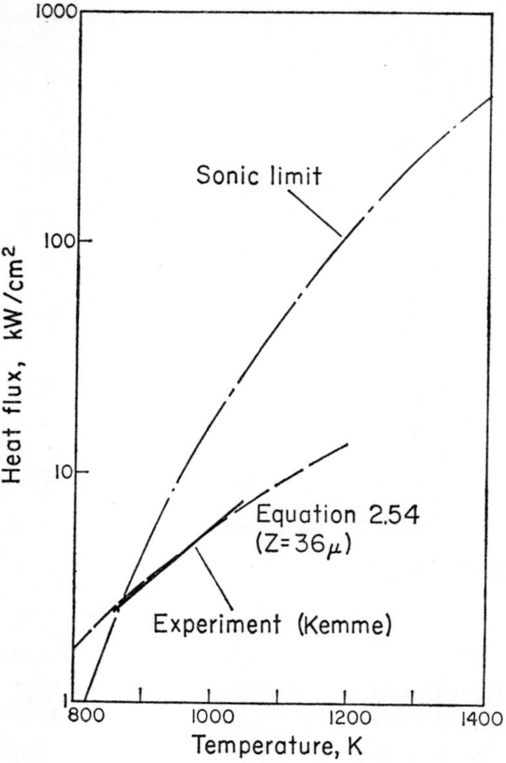

Fig. 2.7. Entrainment and sonic limits of a sodium heat pipe

$$We = \frac{Z.U^2.\rho v}{\sigma} \tag{2.52}$$

where Z is a characteristic dimension of the vapour–liquid interface.

The axial heat flux is

$$q = U.\rho v.L \tag{2.53}$$

hence substituting U, the vapour velocity, from equation 2.52 with We = 1 into equation 2.53 the axial heat flux at the onset of entrainment is

$$q = \sqrt{\frac{\rho v.\sigma.L^2}{Z}} \tag{2.54}$$

Fig. 2.7 gives a comparison between experiment and prediction using equation 2.54 with $Z = 36\ \mu$. At the present time it is not possible to predict on a theoretical basis the value of the characteristic dimension Z.

3. Temperature Distributions and Heat Transfer Coefficients

In this chapter, the method of predicting the temperature drop over the vapour is discussed, as are the resistance to heat transfer of the wick and liquid in the evaporator and condenser and the resistance to heat transfer at the interface between the phases.

Notation

A_c	Interfacial surface area between phases in condenser
A_e	Interfacial surface area between phases in evaporator
C_p	Specific heat of liquid
D_e	Characteristic dimension (4 $\times$ hydraulic radius)
D_p	Particle diameter (metre)
G	Mass flow rate of vapour per unit of free area of wick
g	Acceleration of gravity
h	Heat transfer coefficient
k_L	Thermal conductivity of liquid
k_m	Thermal conductivity of wick material
k_w	Effective thermal conductivity of saturated wick
L	Latent heat of evaporation
M	Molecular weight
$\dot{m}$	Mass flow rate
p	Pressure
p_{Sc}	Vapour pressure corresponding to temperature T_{Sc}
p_{Se}	Vapour pressure corresponding to temperature T_{Se}
p_{Vc}	Vapour pressure corresponding to temperature T_{Vc}
p_{Ve}	Vapour pressure corresponding to temperature T_{Ve}

R	Universal gas constant
r_n	Nucleating site radius
T_{sat}	Saturation temperature
T_{Sc}	Temperature of liquid surface in condenser
T_{Se}	Temperature of liquid surface in evaporator
T_{Vc}	Temperature of vapour in condenser
T_{Ve}	Temperature of vapour in evaporator
ΔT_{sup}	Degree of superheat in liquid at wall
ε	Void fraction of wick
μ_L	Absolute viscosity of liquid
ρ_L	Density of liquid
ρ_V	Density of vapour
σ	Surface tension
ϕ	Condensation coefficient

3.1 AXIAL TEMPERATURE VARIATION

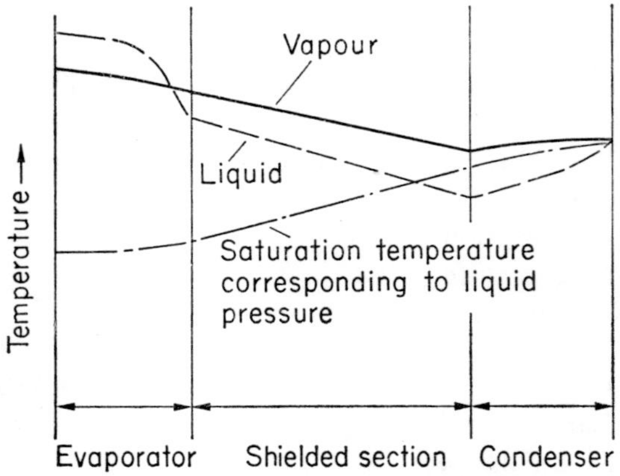

Fig. 3.1. Variation of fluid temperature in heat pipe

Fig. 3.1 illustrates diagrammatically the variation of temperature of the fluid as it circulates round the heat pipe. Starting at the end of the condenser and following the direction of flow, the bulk liquid temperature falls till the liquid reaches the exit from the condenser. In the shielded section, the liquid temperature rises owing to heat transfer from the vapour. In the evaporator, the liquid temperature rises quickly with heat transfer from the wall, until the bulk liquid temperature is above the vapour temperature. As the pressure of the liquid in the evaporator is below that of the vapour, owing to capillary forces, the saturation temperature of the liquid is below the vapour saturation temperature; thus, in the evaporator the liquid is at a temperature above its saturation temperature.

Following now the vapour flow from the end of the evaporator, the temperature drops along the flow path with falling pressure and heat transfer to the liquid, a small part of the vapour condensing, until the inlet to the condenser is reached. In the condensing section there can be a slight temperature recovery associated with a pressure rise in the vapour.

The pressure change over the vapour phase may be evaluated using the methods described in Chapter 2, and the temperature change along the heat pipe can then be obtained from a vapour pressure–temperature curve or Clapeyron's equation. As the pressure change along the vapour path is normally small, the temperature gradient in the vapour phase is also normally small in absolute terms (<2 K), and also in relation to the temperature differences which can occur over the wick and wall in both the evaporator and condenser.

3.2 HEAT TRANSFER IN THE CONDENSER

In a number of investigations, rather large temperature differences have been measured between the vapour and wall in the condenser. Lyman and Huang have analysed this situation in detail, using as a basis the assumption that the heat is transferred by conduction through the water-filled wick, the effective wick thermal conductivity being evaluated from

$$k_{\mathrm{w}} = \varepsilon k_{\mathrm{L}} + (1 - \varepsilon)\, k_{\mathrm{m}} \tag{3.1}$$

At the present time, this equation is recommended as a design basis, but it will tend to give low estimates at higher heat fluxes where some convection occurs. Gorring and

Churchill discuss in more detail methods for estimating the thermal conductivity of heterogeneous materials.

The presence of non-condensable gases in the vapour will radically influence the performance of the condenser, as the gases will be swept into the condenser by the motion of the vapour. In condensing, the vapour must diffuse through the gases to reach the cooled surface, and this creates a considerable temperature gradient in the gas–vapour mixture. Provan has discussed this in some detail.

Fig. 3.2 shows the wall temperature distribution obtained by Schwartz in testing a stainless steel heat pipe with water as the working fluid. During the first few days of operation, the reaction of the water with the stainless steel produces hydrogen with a resultant large drop in the temperature of the condenser wall.

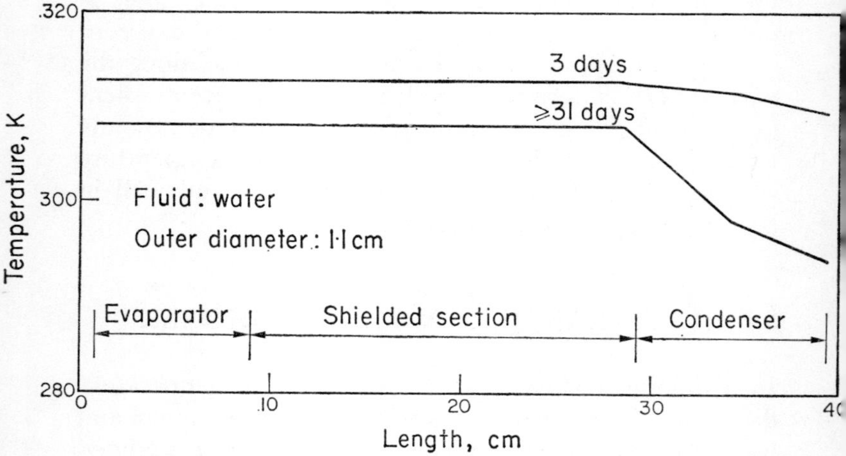

Fig. 3.2. Temperature variation along heat pipe wall including effect of time

3.3 HEAT TRANSFER IN THE EVAPORATOR

Fig. 3.3 illustrates three possible mechanisms of heat transfer in the evaporator. At relatively low rates of heat transfer, the liquid evaporates at a miniscus near the wick surface as shown in Fig. 3.3a. At higher rates of heat transfer, boiling can occur within the wick; this will restrict circulation unless there is adequate provision for the vapour to escape. At yet higher rates of heat transfer, part of the wick runs dry. This has two serious effects: first, very low rates of heat transfer then occur, and second the maximum capillary pressure difference which can be exerted is now associated with pores with their main axis parallel to the wall. This latter point can be important in the case of composite wick structures. Savage has examined numerically situations of the type shown in Fig. 3.3 for the case where the heat is transferred across the end plate of the evaporating section.

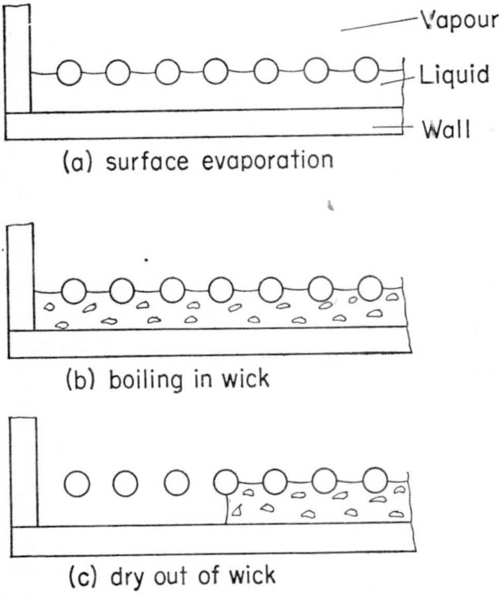

Fig. 3.3. Possible heat transfer mechanisms in evaporator

For the situation illustrated in Fig. 3.3a, where evaporation occurs at the surface of the wick, the effective wick thermal conductivity can be evaluated using equation 3.1. Phillips and Hinderman found that assuming the wick material to have the same thermal conductivity as the liquid gave good agreement for a number of their experiments. The assumption does not hold for conditions close to the onset of nucleate boiling; this is illustrated in Fig. 3.4. In their tests, water was used, with a wick of 0·127 cm thick nickel foam covered by a 200 mesh screen, the total thickness being 0·143 cm. Schwartz found, however, that the effective thermal conductivity was greater than that of water alone. As already mentioned, the use of equation 3.1 will tend to underestimate the heat transfer coefficient at higher temperature difference, as it ignores natural convection effects.

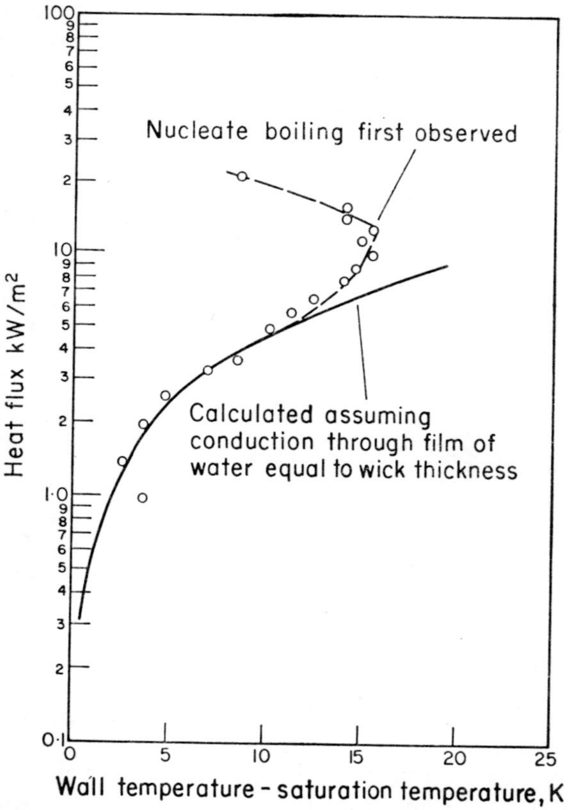

Fig. 3.4. Heat flux to evaporator as a function of liquid superheat at wall

An approximate relationship for the superheat at the wall to give nucleate boiling[(G10)] is

$$\Delta T_{sup} = \frac{2.\sigma.T_{sat}}{L.\rho v.r_n} \tag{3.2}$$

The use of this equation requires knowledge of the nucleating site radius r_n; this is not normally known, however Marto and Mosteller found that a value of $r_n = 0{\cdot}0013$ cm gave good agreement with their observation for the onset of nucleate boiling in a wick with an effective capillary radius of 0·007 cm.

Fig. 3.5a compares heat transfer coefficients obtained under conditions where nucleate boiling occurs or was assumed to occur. Curves shown are obtained by the following procedures:

1. The nucleate boiling correlation of Gilmour for smooth surfaces
2. The empirical curve of Jakob and Fritz for boiling on a grooved surface
3. An equation by Allingham and McEntire obtained by studying heat transfer coefficients for a 2·54 cm diameter copper tube embedded in a water-saturated ceramic fibre-wick material with heat fluxes in the range 3 kW/m² to 30 kW/m².

The equation is

$$\left[\frac{h}{C_p.G}\right]\left[\frac{C_p.\mu_L}{k_L}\right]^{0\cdot 6}\left[\frac{g.\rho_L.\sigma}{p^2}\right]^{0\cdot 21} = 0{\cdot}072\left[\frac{D_e.G}{\mu_L}\right]^{-0\cdot 77} \tag{3·3}$$

Fig. 3.5b extends the comparison to include:
4. The equation developed by Anand[(3)]

$$\left[\frac{h}{C_p.G}\right]\left[\frac{C_p.\mu_L}{k_L}\right]^{0\cdot 6}\left[\frac{g.\rho_L.\sigma}{p^2}\right]^{0\cdot 2} = 0{\cdot}00051\left[\frac{D_e.G}{\mu_L}\right]^{-1\cdot 43} \tag{3.4}$$

for heat transfer to a 1·9 cm outer diameter stainless steel pipe with a wick of 100 mesh stainless steel screen. It is not certain that boiling occurred in these tests.

5. The heat transfer coefficients corresponding to the curve in Fig. 3.4 above the point where nucleate boiling was observed to begin.
6. The empirical curves of Marto and Mosteller for a wick of four layers of 100 mesh plain weave of wire diameter 0·0114 cm. The fluid was water at the pressures indicated.

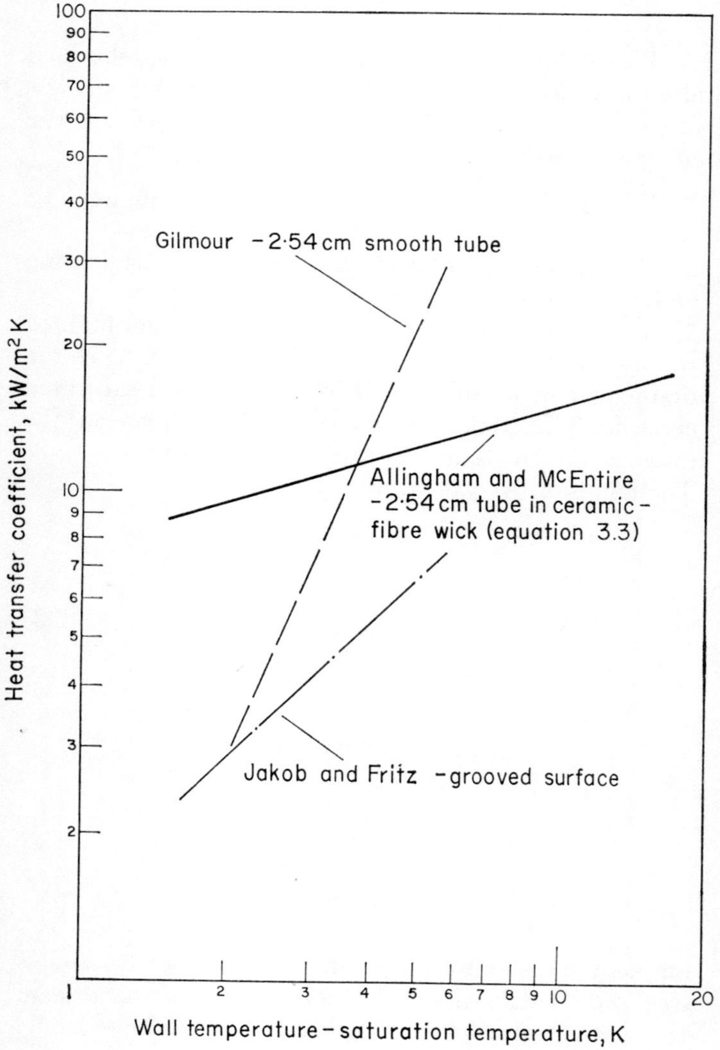

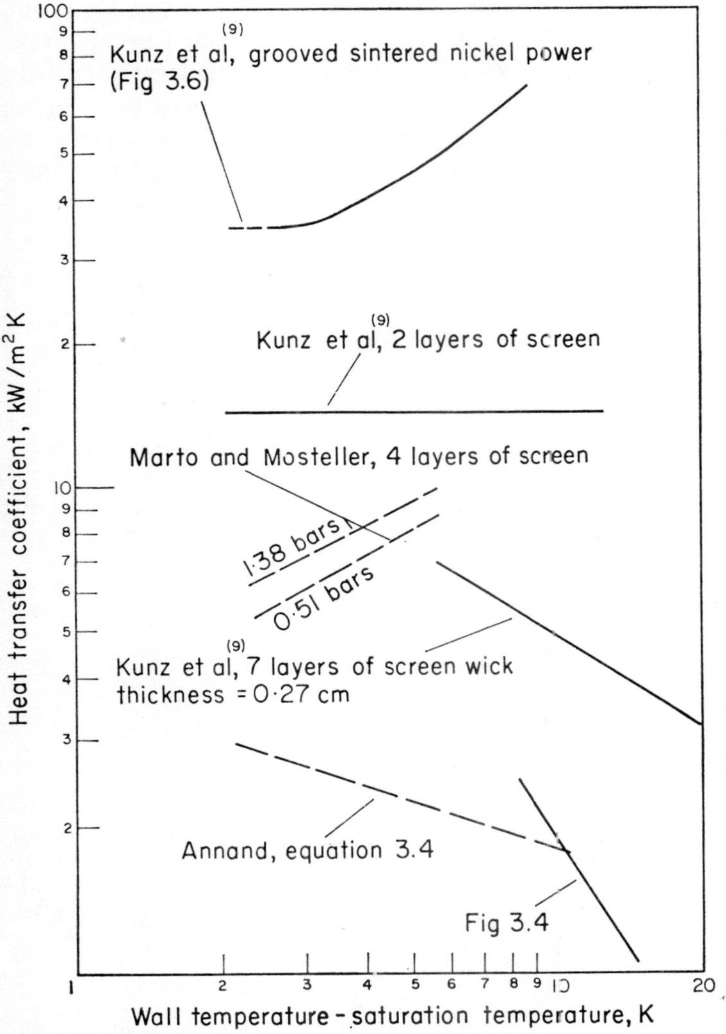

Fig. 3.5. Comparison of heat transfer coefficients for nucleate boiling

7. Some of the extensive data of Kunz *et al.*[9] obtained using water as the fluid at pressures close to atmospheric.

These latter tests indicate the considerable improvements to be obtained in using a wick of two layers instead of seven layers. This is presumably due to the ease with which the vapour may leave the wick in the former case. It also partly explains the high heat transfer coefficients obtained with a wick of sintered nickel powder, grooved to provide ready venting of the vapour. This is further illustrated in Fig. 3.6, where the heat fluxes for a grooved and ungrooved wick of sintered nickel powder are compared.

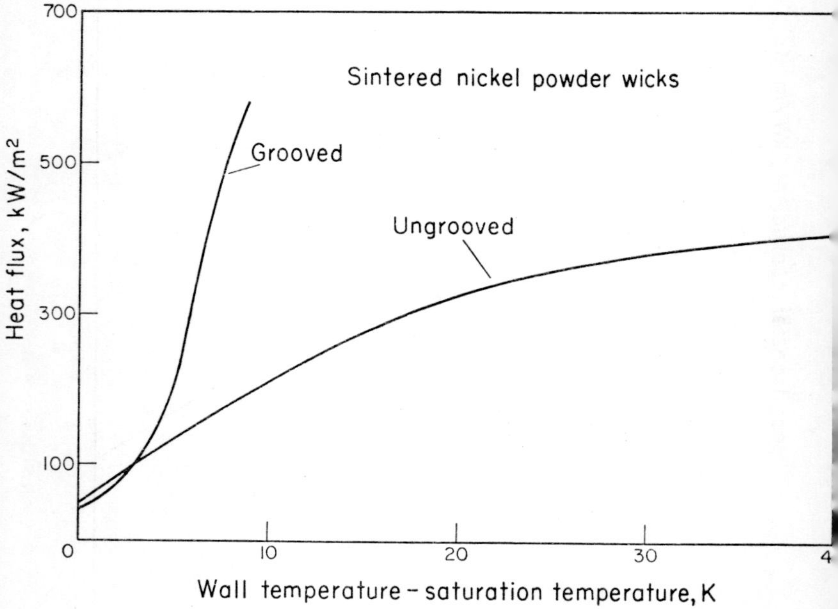

Fig. 3.6. Influence of venting grooves on the temperature differences for given heat fluxes

It is not yet possible to allow for all the factors which influence the heat transfer coefficient, many of which are examined by Kunz *et al.*[9] For certain geometries, however, a good understanding of the heat transfer mechanism has been obtained. Ferrel and Alleavitch, who examined wicks of beads, found that the heat transfer coefficient was independent of the temperature difference. The heat transfer coefficient was a function of the bead diameter only, being given by

$$h = \frac{0{\cdot}00536}{D_p} \tag{3.5}$$

The units of h, the heat transfer coefficient, are $kW/m^2\ K$, and D_p the particle diameter is in metres. Ferrel and Alleavitch develop a satisfactory model to explain this, based on the concept of a thin layer of liquid at the wall with the liquid vaporizing at the interface which is located at the minimum pore diameter in the first layer of particles on the surface.

In practice, it should be possible to attain a coefficient at least as high as given by equation 3.3; this equation is therefore recommended as a conservative basis of design.

At low temperature differences, the wick surfaces give greatly improved heat transfer characteristics as compared with smooth surfaces, and tubes with prepared surfaces are now available commercially for use in conventional heat exchangers to take advantage of this. The Union Carbide Corporation market tubes with porous surfaces having a thickness between 0·1 and 1 mm with pore sizes in the range 1 to 150 microns. Laboratory data produced by Union Carbide quote coefficients similar to those for a grooved wick in Fig. 3.5b; for example, where the temperature difference between wall and vapour is 5·5 K, heat transfer coefficients approaching 57 $kW/m^2\ K$ are quoted.

The burn-out or dry-out heat flux will depend greatly on the geometry of the wick. Ferrel and Alleavitch have developed a model which gives good agreement with their experiments. Costello and Frea found that in their arrangement the wick increased the burn-out flux by about 20 per cent compared with a smooth tube.

3.4 INTERFACIAL RESISTANCE TO HEAT TRANSFER

During surface evaporation in the evaporator, the rate

of evaporation per unit of surface area[C8, K2] is given approximately by

$$\frac{\dot{m}}{A_e} = \phi\sqrt{\frac{M}{2.\pi.R.T_{Ve}}}\left[p_{Se}\left(\frac{T_{Ve}}{T_{Se}}\right)^{1/2} - p_{Ve}\right] \quad (3.6)$$

and similarly in the condenser

$$\frac{\dot{m}}{A_c} = \phi\sqrt{\frac{M}{2.\pi.R.T_{Sc}}}\left[p_{Vc}\left(\frac{T_{Sc}}{T_{Vc}}\right)^{1/2} - p_{Sc}\right] \quad (3.7)$$

Normally the pressure and temperature differences are extremely small. It is only when the vapour pressure is very low that this interfacial resistance is of importance; for example, p_{Vc} the vapour pressure in the condenser may be so small that even with a zero value for the vapour pressure corresponding to the liquid surface temperature (p_{Sc}) the required condensation rate may be unattainable.

4. Selection of the Working Fluid

The working fluid should possess the following properties:

1. High surface tension to provide satisfactory capillary pumping.
2. Good wetting characteristics, for the same reason.
3. Low viscosity to aid pumping.
4. High latent heat of vaporization to aid axial heat transfer.
5. High thermal conductivity to aid heat transfer between fluid, wall and wick.
6. Freezing and boiling points compatible with the range of operation.
7. High density to reduce flow resistance.
8. Compatibility in relation to corrosion with container wall and wick.
9. Chemical stability.

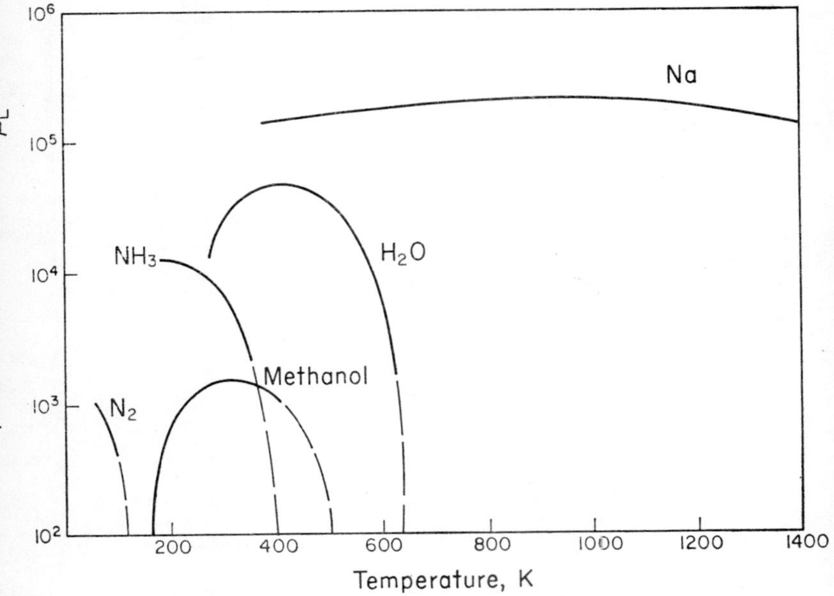

Fig. 4.1. Liquid transport factor for a number of working fluids

Equation 2.11 contains a fluid property group which is referred to as the liquid transport factor,(F2, K2) as the higher the value of this factor the higher the circulation rate in the heat pipe. The liquid transport factor

$$N_e = \frac{\sigma . \rho_L . L}{\mu_L}$$

$$= \frac{\text{(surface tension)} \times \text{(latent heat per unit volume)}}{\text{absolute viscosity}} \tag{4.1}$$

Fig. 4.1 shows this group as a function of temperature for a number of heat pipe working fluids. A rather similar group which includes the vapour properties, developed by Grover, Bohdansky and Busse, is applicable to the particular case of an optimized system with a wick of channels.

Table 4.1 compares the relative values of surface tension of some working fluids, and Table 4.2 gives values(L1) of the contact angle for water and water vapour in contact with various materials. Impurities and the condition of the wall influence the contact angle; this is illustrated in Fig. 4.2.

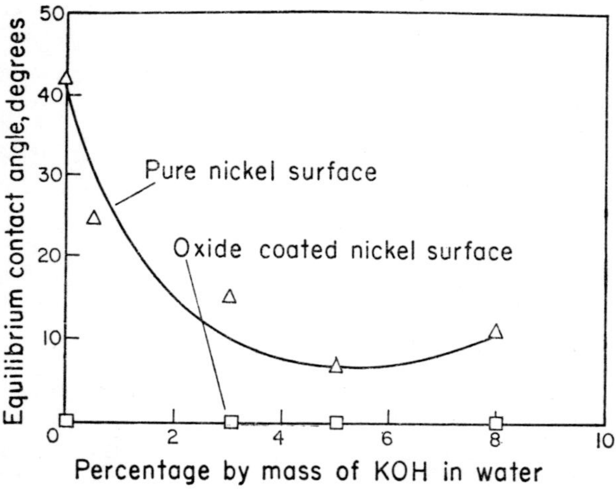

Fig. 4.2. Equilibrium contact angle for aqueous KOH for two nickel surfaces

Table 4.1: Surface tension of fluids: values at atmospheric pressure relative to water at 273 K

SOURCE	TEMPERATURE K	RELATIVE SURFACE TENSION	FLUID
Water	273 (melting point)	1·0	Frank(17)
	300	0·95	
	350	0·83	
	373 (boiling point)	0·78	
Cesium	302 (m.p.)	1·01	Frank(17)
	500	0·87	
	700	0·73	
	956 (b.p.)	0·55	
Potassium	337 (m.p.)	1·14	Watts(5)
	500	0·95	
	900	0·83	
	1033 (b.p.)	0·70	
Sodium	371 (m.p.)	2·52	Frank(17)
	600	2·22	
	900	1·83	
	1161 (b.p.)	1·50	
Lithium	1000	4·56	Watts(5)
	1200	4·26	
	1590 (b.p.)	3·70	
Silver	1234	12·2	Watts(5)
	1500	11·5	
	2000	10·7	
	2450 (b.p.)	9·8	

Table 4.2. Contact angles for water and its vapour with various substances (Reference L1)

SUBSTANCE	CONTACT ANGLE-DEGREES		
	ADVANCING		RETREATING
Graphite	55–60		59
Graphite		17·8–28·5	
Quartz	0–10		0
Glass		0	
Glass		$\cos \alpha = 0{\cdot}694$	
Glass	39·5		33
Iron		$\cos \alpha = 1{\cdot}000$	
Nickel		$\cos \alpha = 0{\cdot}994$	
Nickel		27°	
Copper		$\cos \alpha = 0{\cdot}998$	
Copper	2·1–45		0
Zinc	1·8–15·7		0
Aluminium		$\cos \alpha = 0{\cdot}997$	

Table 4.3 compares the latent heats of a number of fluids. The latent heat should be as high as possible, as the mass flow rate to transport a given quantity of heat is inverseley proportional to the latent heat (equation 1.2).

The vapour pressure should not be so high that the container requires excessively thick walls, but at the same time too low a density can result in too high a flow resistance or the onset of choking (Chapter 2.6). Too low a vapour pressure can also result in too great a resistance to heat transfer at the interface between the vapour and liquid (Chapter 3.4). The vapour pressure curves of several working fluids are given in Fig. 4.3.

Where possible, a fluid should be used in a region where the gradient of the vapour pressure–temperature curve is high,[P2] so that a pressure change will cause only a small temperature change. From this viewpoint, potassium is preferable to sodium at 950 K as illustrated in Fig. 4.4.

Compatibility between the working fluid and the container and wick is essential from a corrosion viewpoint. Corrosion of the wall can retard heat transfer and block the pores of the wick. Also, any chemical reaction producing gases must be avoided, as the gases will move to the condenser and impede heat transfer (Chapter 3.2).

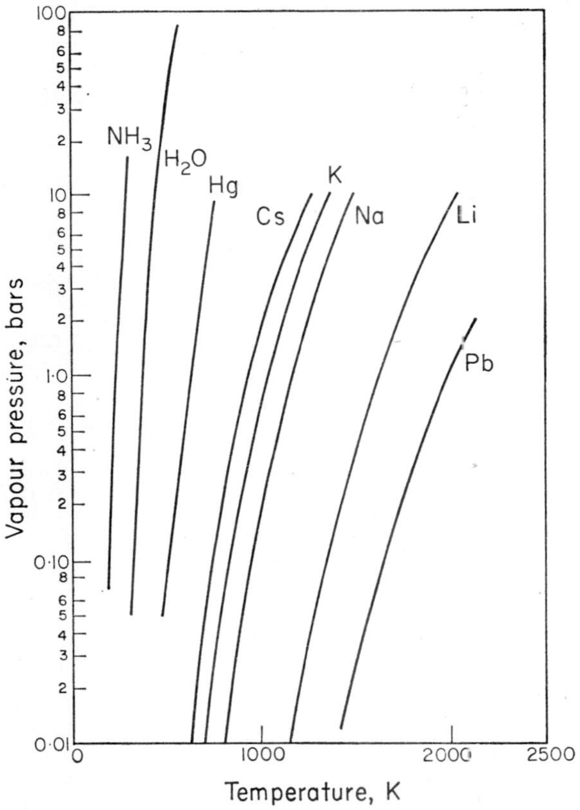

Fig. 4.3. Vapour pressure as function of temperature for various working fluids

Water and aluminium have been found(S6) to react to give hydrogen, as have water and stainless steel. Ammonia and acetone are compatible with aluminium and ammonia with stainless steel. Compatibility is discussed further in Chapter 7.

Table 4.3 Latent heat of fluids: Values relative to water at 273 K (2500 kJ/kg)

FLUID	TEMPERATURE K	RELATIVE LATENT HEAT	SOURCE
Ammonia	196	0·59	
	211	0·58	Ashrae
	228	0·56	Handbook
	240	0·55	
Water	273	1·0	Frank(17)
	300	0·97	
	350	0·92	
	373	0·90	
Cesium	302	0·23	Frank(17)
	500	0·22	
	700	0·21	
	956	0·20	
Potassium	337	0·90	Watts(5)
	500	0·87	
	900	0·79	
	1033	0·77	
Sodium	371	1·90	Frank(5)
	600	1·77	
	900	1·62	
	1161	1·52	
Lithium	452	—	Watts(5)
	1000	8·7	
	1200	8·4	
	1590	7·7	
Silver	1234	—	Watts(5)
	1500	0·97	
	2000	0·96	
	2450	0·94	

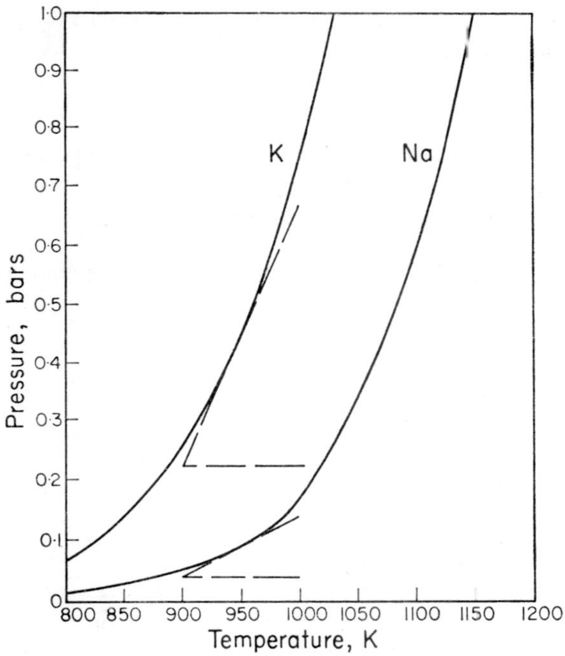

Fig. 4.4. Comparison of vapour pressures of sodium and potassium

5. Characteristics of Wicks

The performance of a heat pipe is critically dependent on the performance of the wick structure used to return the liquid from the condenser to the evaporator.

Aspects of the wick and its performance discussed in the chapter are

1. The geometric arrangement of the wick.
2. The methods for determining the resistance to flow through the wick.
3. The method of determining the capillary rise or capillary driving pressure.

Notation

A Cross-section of wick
A_c Cross-section of water column (Fig. 5.2)
A_L Cross-section of wick occupied by liquid
A_s Cross-section of wick not occupied by wick material
a Width of channel, equation 5.1
b Breadth of channel, equation 5.1
D Pipe diameter
D_p Particle diameter
g Acceleration of gravity
h Height of water column (Fig. 5.2)
K Permeability (specific, equation 5.7)
k Permeability (as defined by Darcy, equation 5.12)
l Length of wick occupied by liquid
l_e Equilibrium capillary rise
$\dot{m}$ Mass flow rate
N Number of channels (equation 5.1)
Δp_c Pressure difference due surface tension
Δp_G Pressure difference due gravity
Δp_L Pressure difference due to friction
R Radius of capillary passage
Re Reynolds number
r Radius of curvature of liquid surface
t Time interval
V Volume of liquid (equation 5.41)

Z Radial clearance (equation 5.2)
α Contact angle
ε Void fraction (free volume/total volume of wick)
θ Inclination of wick to horizontal
μ Absolute viscosity of liquid
ρ_L Density of liquid
σ Contact angle
ϕ Correction factor in equation 5.1.

5.1 GEOMETRIC ARRANGEMENT

The original wick structures used in heat pipes were materials such as woven cloth, fibre glass, porous metal and wire mesh. These wicks will be referred to as homogeneous wicks, to distinguish them from wicks made up of combinations of these materials; the latter will be referred to as composite wicks.

Fig. 5.1a illustrates diagrammatically the use of homogeneous wicks. The wick is attached to the wall of the heat pipe in such a manner as to ensure good contact between the wick and the wall in the regions where heat transfer occurs. Good contact helps in maintaining satisfactory heat transfer rates from and to the heat pipe fluid.

Channels in the wall, as illustrated in Fig. 5.1b, have also been used; a further development is to cover the channel with a fine screen as illustrated in Fig. 5.1c. One advantage of this arrangement is that it reduces entrainment of the liquid flowing in the wick by the vapour flowing from the evaporator section of the heat pipe to the condensing section. More important, the screen can be of a very small pore size to give a high capillary driving pressure, without significantly increasing the flow resistance in the channels. Kemme[3] tested the arrangements shown in Figs 5.1b and c and found a threefold improvement in the performance of a heat pipe with the latter arrangement.

Similarly, a wire screen can be used in conjunction with a homogeneous mesh as illustrated in Fig. 5.1d. The mesh adjacent to the wall must have a pore size sufficient to ensure start-up of the heat pipe, but the pore size governing the capillary driving pressure during steady-state operation is that of the wire screen, which can therefore be made small without influencing the resistance to flow as significantly as in the case of a homogeneous wick.

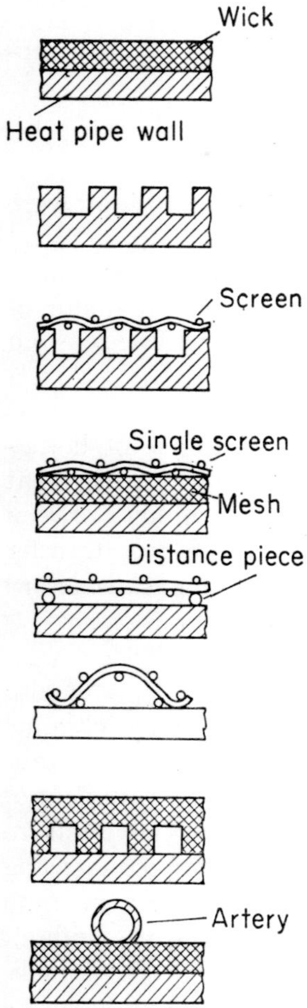

Fig. 5.1a. Homogeneous wick

Fig. 5.1b. Channels

Fig. 5.1c. Channels and screen

Fig. 5.1d. Mesh and screen

Fig. 5.1e. Screen and gap. Screen and annulus (in tubular heat pipes)

Fig. 5.1f. Corrugated screen

Fig. 5.1g. Channelled wick

Fig. 5.1h. Artery

Fig. 5.1. Geometric arrangement of wicks

Another arrangement uses a wire screen, as shown in Fig. 5.1e. The mesh is located a small distance from the heat pipe wall by means of distance pieces, which can be fine wires running along the direction of flow. The distance between the wall and mesh is sufficiently small to provide adequate capillary forces for the initial wetting of the heat pipe; the steady-state capillary driving pressure is again the wire screen. The use of distance pieces can be avoided by using a corrugated screen, as shown in Fig. 5.1f.

Where porous metal is used, an alternative arrangement to that of Fig. 5.1c may be used, as shown in Fig. 5.1g.

Fig. 5.1h illustrates the use of an "artery" to provide a low resistance path to the liquid returning to the evaporation section. This system is useful if boiling is otherwise likely to occur within the wick during the return of the liquid to the evaporator. It has the disadvantage, however, that the system requires an additional capillary network to distribute the liquid over the heat transfer surfaces. Katzoff[1] describes a heat pipe utilizing this principle.

5.2 RESISTANCE TO LIQUID FLOW

Certain of these geometries are such that the flow resistances can be estimated by relatively simple formulae. The formulae below correspond to laminar flow, which is the normal flow condition encountered in heat pipes.

1. *Channels with Screen (Fig. 5.1)*
Treating the screen as a smooth surface[M2] gives

$$\Delta p_L = -\frac{4\mu \dot{m} l}{a^2 b^2 \rho_L N \phi} \tag{5.1}$$

The correction factor ϕ is given in Table 5.1.

Equation 5.1 can be obtained analytically. The case of the open channel is a more complex situation. Formulae for that situation are presented by Andel and Busse[6].

2. *Annulus (Fig. 5.1)*

$$\Delta p_L = -\frac{12\mu \dot{m} l}{\pi D Z^3 \rho_L} \tag{5.2}$$

Table 5.1. Correction factor ϕ (equation 5.1 as function of a/b)

a/b	1·0	0·8	0·6	0·4	0·2	0
ϕ	0·141	0·138	0·124	0·1	0·058	0

In the case of a circular heat pipe it is advantageous to have the annulus eccentric, for if the screen should touch the wall at one point the flow is increased 2·5 times relative to the concentric case. Thus, for the eccentric or crescent annulus, with wall contact,

$$\Delta p_L = -\frac{4{\cdot}8\mu\dot{m}l}{\pi D Z^3 \rho_L} \tag{5.3}$$

where Z the radial clearance is evaluated on the basis of the annulus being concentric.

3. *Artery (Fig. 5.1h)*
The normal formula(M2) for laminar flow in circular tubes applies

$$\Delta p_L = -\frac{40{\cdot}75\mu\dot{m}l}{D^4 \rho_L} \tag{5.4}$$

4. *Layer of Wire Screens (Fig. 5.1a)*
Schmidt(4) has measured the pressure drop through a wick made up of a layer of screens and obtained the following equation

$$\Delta p_L = -\left(\frac{4400}{\mathrm{Re}} + 25\right)\frac{1-\varepsilon}{\varepsilon^3}\frac{\dot{m}^2}{A^2 D_p}\frac{1}{\rho_L} \tag{5.5}$$

where ε is the wick porosity (void volume/total volume)
D_p is the pore diameter (six times the screen wire)
A is the total cross-section of the wick in the direction of flow and
Re is the Reynolds number defined by

$$\mathrm{Re} = \frac{D_p\dot{m}}{\mu A}\frac{1}{1-\varepsilon} \tag{5.6}$$

The experiments covered a range of Reynolds numbers from 2 to 600. It appears likely that turbulent flow was occurring at the higher Reynolds numbers.

5.3 PERMEABILITY

During laminar flow through a porous material, the pressure difference due to friction is given by Darcy's equation

$$\Delta p_L = -\frac{\mu \dot{m} l}{\rho_L K A} \tag{5.7}$$

where in this form K is the specific permeability. Equation 5.7 can be considered as a general statement of equations 5.1 to 5.3.

The permeability can be measured readily using a rig such as shown(F2) in Fig. 5.2. The pressure drop over the wick is obtained from the head of liquid, and the liquid flow rate is obtained by volumetric or gravimetric means from the collecting vessel.

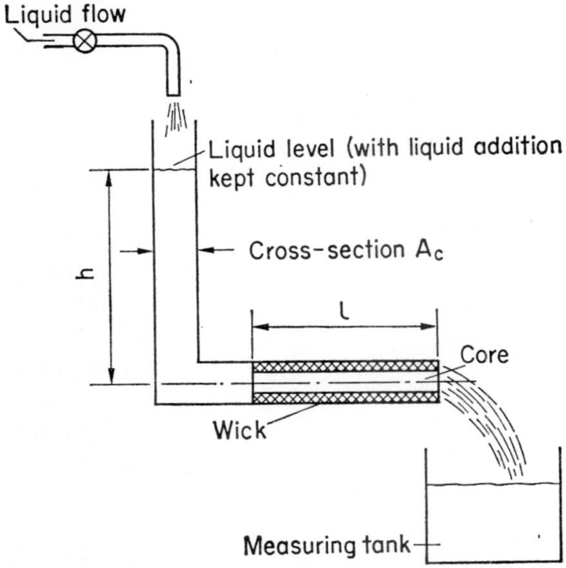

Fig. 5.2. Rig for measuring permeability

The method of attaching the wick in the annulus can have a significant effect on the measurement, owing to the influence of any slight gap between the wick and the walls, and to the influence of compaction of the wick in the case of non-rigid materials. These effects cause[F2] as much as a hundredfold variation in observed permeability values. In section 5.7 more suitable methods for determining the permeability of non-rigid wicks are discussed.

Cosgrove,[10] for example, used a transient method of determining the permeability. The liquid supply to the liquid column (Fig. 5.2) was stopped, and the rate of fall of the liquid column then measured. Note that

$$\Delta p_L = -h\rho_L g \tag{5.8}$$

and

$$\frac{\dot{m}}{A_c \rho_L} = \frac{dh}{dt} \tag{5.9}$$

where A_c is the cross-section of the liquid column. Hence from equations 5.7 to 5.9 the permeability is

$$K = \frac{\mu}{\rho_L g} \frac{A_c}{A} \frac{dh}{dt} \frac{l}{h} \tag{5.10}$$

On integrating

$$K = \frac{\mu}{\rho_L g} \frac{A_c}{A} \frac{l}{t} \ln \frac{h_{initial}}{h_{final}} \tag{5.11}$$

where t is the time for the head to change from $h_{initial}$ to h_{final}.

The permeability K is a function of the geometry of the passages of the wick, but independent for laminar flow on flow rate and liquid properties. Variations in K with time have been observed[C10, L2] in a heat pipe but in one case at least this has been demonstrated to be due to air bubbles becoming trapped in the wick. Deposition of suspended solid matter in the pores of the wick is another reason for variations with time.

For the particular case of a wick consisting of a bed of uniform spheres, the permeability can be predicted[C10, F8] using the Blake–Kozeny relationship

$$K = \frac{D_p^2 \varepsilon^2}{150(1 - \varepsilon)^2} \tag{5.12}$$

However, analytic means of predicting K are not normally available.

It should be noted that the symbol K is referred to as the "specific" permeability, to distinguish it from the permeability as defined by the original statement of Darcy's equation or law, which was written

$$\Delta p_L = -\frac{\dot{m} . l}{\rho_L . k . A} \tag{5.13}$$

Thus

$$K = k\mu \tag{5.14}$$

It is normal now to refer to K as the permeability, and this will be the practice throughout this book. It is also important to note that in much of the recent literature the symbol K is used for the reciprocal of the permeability.

Typical values for the permeability of a number of materials are given in Table 5.2. The table also includes values of the capillary rise, which is now discussed in some detail.

5.4 CAPILLARY RISE

The "capillary rise" of a fluid in a wick is the maximum height to which the fluid will rise vertically, as shown in Fig. 5.3 (where, for simplicity, a tube is illustrated).

In a single capillary, the curvature r of the interfacial surface creates a pressure difference across the interface of

$$\Delta p_c = \frac{2\sigma}{r} \tag{5.15}$$

or if the capillary is not circular

$$\Delta p_c = \sigma\left(\frac{1}{r_1} + \frac{1}{r_2}\right) \tag{5.16}$$

Table 5.2. Characteristics of wicks

SAMPLE	MEDIAN PORE SIZE microns	POROSITY %	CAPILLARY RISE cm	PERMEABILITY m^2	FLUID	SOURCE
Layer of nickel screens	>100	62·5	4·85	$6{\cdot}65 \times 10^{-10}$	Water at 75°F	Kunz *et al*[9]
	50	67·6	23·4	$0{\cdot}776 \times 10^{-10}$	Water at 75°F	Kunz *et al*[9]
Nickel powder	>70	65·8	24·6	$2{\cdot}73 \times 10^{-10}$	Water at 75°F	Kunz *et al*[9]
Felted fibres	43·8	86·8	>40·3	$0{\cdot}437 \times 10^{-10}$	Water at 75°F	Kunz *et al*[9]
	43·8	86·8	10·7	$0{\cdot}437 \times 10^{-10}$	Freon 113 at 75°	Kunz *et al*[9]
Silica vitreous fibres	—	—	24·9	$0{\cdot}99 \times 10^{-11}$	Water at 70°F	Ginwala
Fibreglass cloth	—	—	25·4	$6{\cdot}05 \times 10^{-13}$	Water	Feldman [2]
Leather	—	—	—	$0{\cdot}95$ to $1{\cdot}2 \times 10^{-13}$	—	Scheidegger
Cigarette				$1{\cdot}1 \times 10^{-9}$		Scheidegger

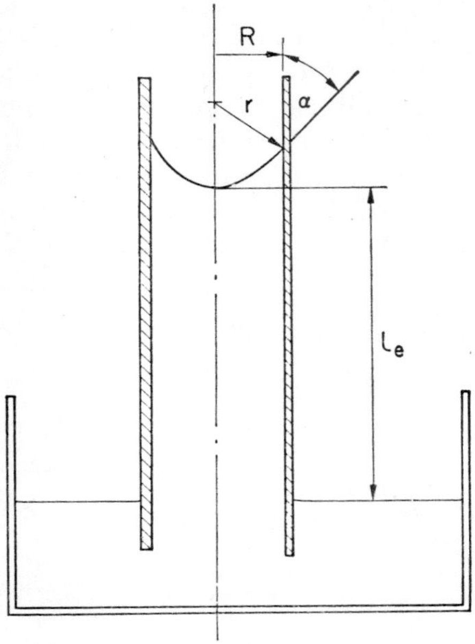

Fig. 5.3. Capillary rise in tube

where r_1 and r_2 are the principal radii of curvature of the miniscus. For convenience the case of circular capillaries will be considered. From geometric consideration

$$\frac{R}{r} = \cos \alpha \tag{5.17}$$

hence substituting equation 5.17 in 5.15

$$\Delta p_c = \frac{2 \cos \alpha}{R} . \sigma \tag{5.18}$$

This pressure difference, when a static equilibrium exists, is balanced by the gravitational force on the column, i.e.

$$\Delta p_c = l_e \rho_L g \tag{5.19}$$

Hence, from equations 5.18 and 5.19, the equilibrium capillary rise is

$$l_e = \frac{2 \cos \alpha}{gR} . \frac{\sigma}{\rho_L} \tag{5.20}$$

In certain cases, such as a tube, it is possible to evaluate the capillary rise theoretically. Another case is that of the randomly packed bed of spheres; a proportion of the spheres will have the closest packing possible (a cubic array) and, as discussed later, these pores (the spaces between the spheres) will be those governing the equilibrium liquid level. For a cubic array, according to Luikov, the effective pore radius is 0·205 times the particle diameter, i.e.

$$r = 0{\cdot}205 D_p \tag{5.21}$$

Hence, from equations 5.20 and 5.21

$$l_e = \frac{2 \cos \alpha}{\rho_L g \; 0{\cdot}205 D_p} . \sigma \tag{5.22}$$

For water at 21°C, assuming $\alpha = 0$, gives

$$l_e = \frac{7{\cdot}26 \times 10^{-5}}{D_p} \tag{5.23}$$

where l_e and D_p are in metres.

As shown in Fig. 5.4, Ferrel and Alleavitch have found this gives good agreement with experiment. Scheidegger gives other geometries for which the effective pore radius is known.

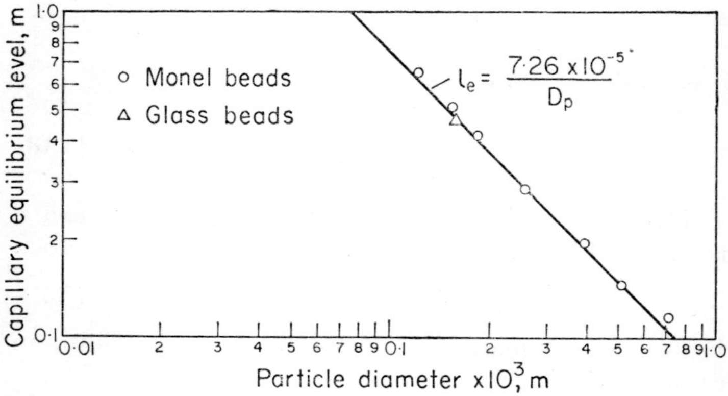

Fig. 5.4. Comparison of measured capillary equilibrium level with theory

However, in many cases, experiment is the only method of determining the capillary rise. Some experimental methods are now described.

5.5 EXPERIMENTAL DETERMINATION OF CAPILLARY RISE

Cosgrove[10] observed visually the rise of liquid in a wick consisting of a bed of particles. As can be seen in Fig. 5.5 the liquid rises very slowly, particularly in the case of beds of small particle size, where it can take over a day to reach equilibrium. Also shown in Fig. 5.5 are data obtained by what can be described as a "falling column" technique. In this case, the wick is immersed initially in a tank of liquid and is slowly withdrawn from the tank, a column of liquid being lifted by the capillary forces in the wick. The equilibrium capillary rise was considered to have been reached when the liquid column broke away from the wick. This technique can be carried out much more quickly than the "rising column" technique.

The "falling column" technique gives higher values of the capillary rise than the "rising column" technique because of capillary hysteresis. Capillary hysteresis can be best described with reference to the cross-section of wick shown in Fig. 5.6. With a rising column, equilibrium is reached when the level is at *A*; with a falling column, equilibrium is reached at *B* where the pore size is smaller than at *A* and is thus capable of supporting a longer column. The pore size between *A* and *B* is too large for capillary forces to raise the liquid from *A* to *B*. This possible variation of liquid level is "capillary hysteresis".

Whether the capillary rise obtained from a falling column or a rising column experiment is more relevant to the case of maximum fluid circulation in a heat pipe depends on a number of operational aspects related to the transient characteristics of the heat pipe. Cosgrove[10, 11] found that the best agreement between theory and experiment for the maximum heat transfer rate in a heat pipe was obtained using capillary rise values obtained by the falling column technique.

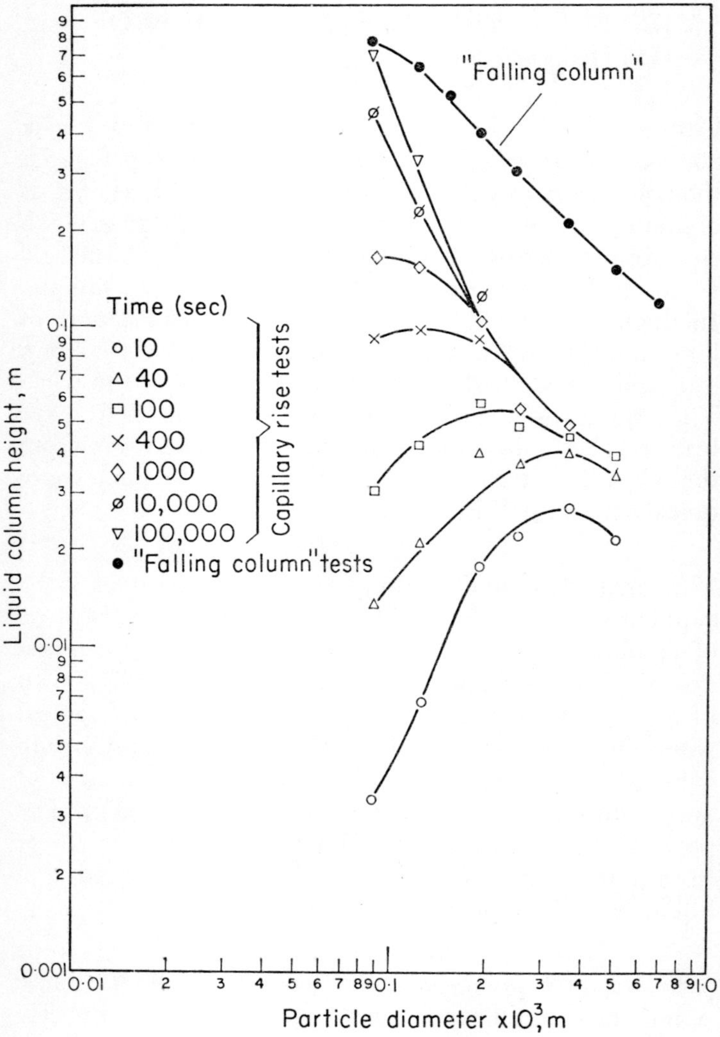

Fig. 5.5. Variation of capillary rise with time and comparison with "falling column" technique

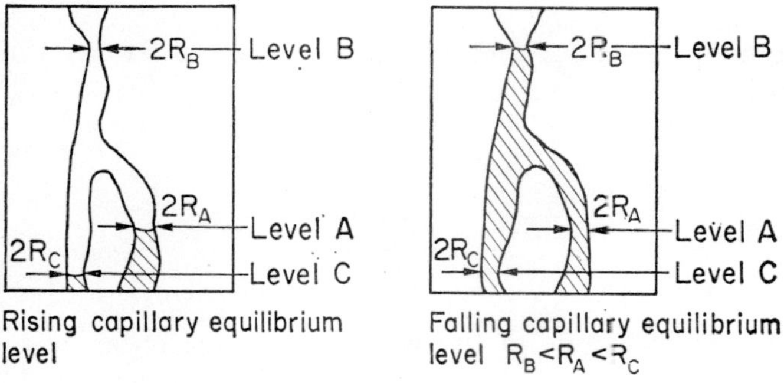

Fig. 5.6. Capillary hysteresis

In determining the liquid level experimentally, it is not always possible to detect the liquid level visually. Farran and Starner used pairs of wire electrodes at discrete points within the wick; when the wick is dry there is a considerable wick resistance to current, but this falls many-fold when the wick becomes wet. Langston and Kunz[(3)] have used litmus paper placed against the side of the wick to determine the liquid level; an aqueous solution of 0·03 per cent potassium hydroxide was used as the liquid.

5.6 DETERMINATION OF THE CAPILLARY DRIVING PRESSURE

In the capillary rise experiments described above, the capillary driving pressure during the transient conditions before equilibrium is reached is balancing the weight of liquid above the tank surface and the pressure drop due to friction over the column length; the forces due to momentum change are negligible.

The mass flow of liquid through the wick is related to the rate of rise of the column in the equation

$$\frac{\dot{m}}{A_L \cdot \rho_L} = \frac{dl}{dt} \tag{5.24}$$

where A_L is the cross-section of the wick occupied by liquid at the liquid front.

From equations 5.7 and 5.24 the pressure difference due to friction is obtained as

$$Ap_{\mathrm{c}} = -\frac{\mu}{K}.\frac{A_{\mathrm{L}}}{A}.l.\frac{dl}{dt} \tag{5.25}$$

and the pressure difference due to gravity is

$$Ap_{\mathrm{G}} = -g.\rho_{\mathrm{L}}.l.\sin\theta \tag{5.26}$$

where θ, the angle of the wick to the horizontal, has a value between 0 and 180° for upward flow of the liquid.

Thus the pressure balance as the liquid rises is

$$\begin{aligned}\Delta p_{\mathrm{c}} &= -\Delta p_{\mathrm{G}} - \Delta p_{\mathrm{L}} \\ &= g.\rho_{\mathrm{L}}.l.\sin\theta + \frac{\mu}{K}.\frac{A_{\mathrm{L}}}{A}.l.\frac{dl}{dt}\end{aligned} \tag{5.27}$$

If the permeability and liquid cross-section are known, the effective capillary pressure difference can be determined by measuring the rate of advance dl/dt of the interface in the wick.

Farran and Starner have used the above technique and also that illustrated in Fig. 5.7 where a steady-state

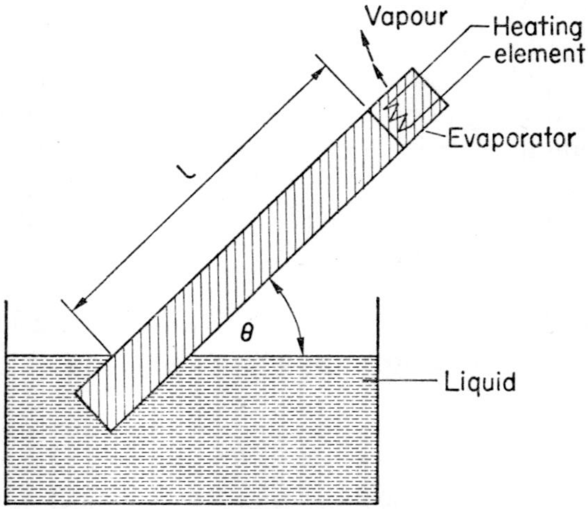

Fig. 5.7. Evaporation test: Farran and Starner

situation is examined. The upper end of the wick is heated to evaporate the liquid rising up the wick, the liquid flow rate being measured by weighing the fluid in the tank.

The pressure balance in this case is

$$\Delta p_c = g.\rho_L.l.\sin\theta + \frac{\mu.\dot{m}.l}{\rho_L.K.A} \tag{5.28}$$

Fig. 5.8 compares flow rates obtained by the latter method with values obtained by the rising interface technique (equation 5.24). For small values of l both techniques give the same flow rate, but diverge with increasing wick length. This arises because the closer the interface approaches the equilibrium level the greater are the number of pore passages unable to exert sufficient capillary driving force; thus, the wick ceases to be completely saturated as was assumed by Farran and Starner in evaluating A_L for use in equation 5.24. This lack of complete saturation of the wick is illustrated in Fig. 5.6; interface C is considerably below interface A as the pore size at C is larger than at A. Between the level of interface A and interface C the wick is not completely saturated. This lack of complete saturation was confirmed by Farran and Starner in a test in which, as the interface reached a height of 51 cm, 5 cm of material was cut from the wick between 48·5 and 53·5 cm and weighed. The average liquid cross-sectional area was then estimated, and it was found that

$$A_L = 0{\cdot}4A_s \tag{5.29}$$

where A_s is the void cross-section. In Fig. 5.7 the mass flow rates from equation 5.24 have been evaluated, assuming

$$A_L = A_s \tag{5.30}$$

when the column height is 51 cm. Using equation 5.29 would give much closer agreement between the two methods of obtaining the mass flow rate.

It is not possible, as a rule, to measure capillary rise and effective capillary driving pressure at the actual condition of heat pipe operation, and values must be corrected to allow for changes in physical properties. For example, if the effective capillary pressure is evaluated at condition 1

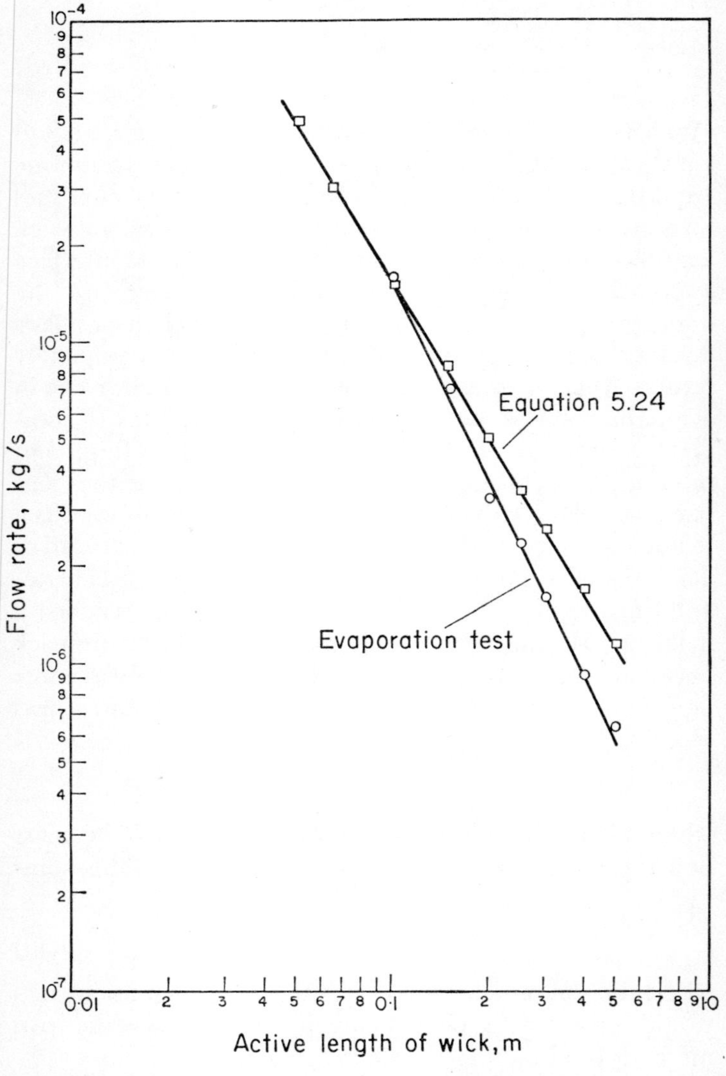

Fig. 5.8. Comparison of flow rates obtained by transient (equation 5.24) and steady-state evaporation tests

and the heat pipe operates at condition 2, then from equation 5.18

$$\Delta p_{c2} = \Delta p_{c1} \cdot \frac{\sigma_2}{\sigma_1} \cdot \frac{\cos \alpha_2}{\cos \alpha_1} \tag{5.31}$$

The subscript indicates the condition at which the property holds. Similarly, from equation 5.20

$$l_{e2} = l_{e1} \frac{\sigma_2 . \cos \alpha_2}{\rho_{L2}} \cdot \frac{\rho_{L1}}{\sigma_1 . \cos \alpha_1} \tag{5.32}$$

In the case of water, with increasing temperature σ and ρ_L will decrease, σ to a greater extent; α might also be expected to decrease if heat is being added due to evaporation. Thus, heating might be expected to decrease capillary rise with water, whereas Cosgrove, Ferrel and Carnesale observed the opposite trend.

5.7 SIMULTANEOUS DETERMINATION OF CAPILLARY PRESSURE AND PERMEABILITY

When heat pipes operate in the horizontal plane or in the absence of gravity, it is often unnecessary to know the capillary pressure and permeability; the product of the capillary pressure and the permeability may be adequate for some design purposes. Equation 5.27 reduces to

$$\Delta p_c . K = \mu . \frac{A_L}{A} . l . \frac{dl}{dt} \tag{5.33}$$

and equation 5.28 to

$$\Delta p_c . K = \mu . l . \frac{\dot{m}}{A . \rho_L} \tag{5.34}$$

Farran and Starner, and Chato and Streckert have demonstrated the use of this approach. Integrating equation 5.33 on the assumption that l and t are the only variables gives

$$\Delta p_c . K = \frac{\mu}{2} . \frac{l^2}{t} . \frac{A_L}{A} \tag{5.35}$$

It has been found[(F1)] that a logarithmic plot of experimental data for l against t does give a straight line of slope 2, demonstrating that $\Delta p_c . K$ is, as expected, a constant.

Somewhat similar approaches can be used when gravity cannot be ignored. Equation 5.27 can be regrouped

$$\frac{dl}{dt} = \frac{\Delta p_c . K . A}{\mu . l . A_L} - \frac{g . \rho_L . \sin\theta . K . A}{\mu . A_L} \qquad (5.36)$$

Thus if dl/dt is plotted to a base of $1/l$ a linear plot should be obtained (if A/A_L or the saturation is not varying), the intersect of the line with the abscissa giving the final term in equation 5.36 and the slope of the line giving the value of the group

$$\frac{Ap_c . K . A}{\mu . A_L}$$

Knowing the final term in equation 5.36 enables the permeability to be evaluated, thus Δp_c is then the only unknown in the above group and can be evaluated also. Hence, from a single experiment in which the advance of the interface is measured, both the capillary pressure and the permeability can be obtained. A similar procedure is to use equation 5.27 as the starting point; in this case $\dot{m}$ is plotted to a base of $1/l$.

Farran and Starner have examined the use of these procedures and have found that non-linear relationships were obtained when dl/dt and $\dot{m}$ are plotted against $1/l$. A possible explanation for this is that the wick was not completely saturated for all lengths of wetted wick; i.e. in equation 5.36 the ratio A/A_L is a function of l. Reasons for this lack of complete saturation in the case of inclined wicks have been discussed above. The lack of complete saturation also influences the liquid pressure gradient, increasing it, as the liquid is no longer flowing through all the pores of the wick.

5.8 CIRCULATION IN THE ABSENCE OF GRAVITY

If the pressure gradients in the vapour are negligible during the operation of a heat pipe in the absence of gravity, then from equation 5.34 the circulation in the heat pipe is given by

$$\dot{m} = \frac{(\Delta p_{\mathrm{c}}\,.\,K)\,.\,A\,.\,\rho_{\mathrm{L}}}{\mu\,.\,l} \tag{5.37}$$

If the group $(\Delta p_{\mathrm{c}} K)$ is determined at a condition other than the operating condition of the heat pipe, then from equation 5.31

$$\dot{m} = \frac{(\Delta p_{\mathrm{c}}\,.\,K)\,.\,A\,.\,\rho_{\mathrm{L}}}{\mu\,.\,l}\,.\,\frac{\sigma_2 \cos \alpha_2}{\sigma_1 \cos \alpha_1} \tag{5.38}$$

It is not necessary to correct for changes in permeability, as it is independent of the fluid properties (if laminar flow is occurring).

An alternative method does not require the evaluation of the group $(\Delta p_{\mathrm{c}} K)$. Combining equations 5.35 and 5.37 gives

$$\dot{m} = \frac{l}{2t}\,.\,A_{\mathrm{L}}\,.\,\rho_{\mathrm{L}} \tag{5.39}$$

Thus, the liquid flow rate for a length l of wick can be evaluated using equation 5.39 by measuring the time t for the liquid front to move along a distance l. The circulation in the heat pipe is then obtained by correcting as indicated in the case of equation 5.38.

$$\dot{m} = \frac{l}{2t}\,.\,A_{\mathrm{L}}\,.\,\rho_{\mathrm{L}}\,.\,\frac{\sigma_2}{\sigma_1}\,.\,\frac{\cos \alpha_2}{\cos \alpha_1} \tag{5.40}$$

Alternatively, the volume of liquid (V) added in time t can be measured. It is evident that

$$V = l\,.\,A_{\mathrm{L}} \tag{5.41}$$

thus, from equations 5.40 and 5.41 the heat pipe circulation would be

$$\dot{m} = \frac{V.\rho_L}{2t}.\frac{\sigma_2}{\sigma_1}.\frac{\cos\alpha_2}{\cos\alpha_1} \qquad (5.42)$$

Chato and Streckert have used this general approach. This latter method has the advantage that the liquid cross-section (A_L) does not require to be known. However, it is nevertheless only applicable when the wick is saturated as otherwise equation 5.35 does not hold.

6. Start-up and Control

6.1 START-UP

Where the wick in initially saturated, as illustrated diagrammatically in Fig. 1.4, the addition of heat to the liquid in the evaporator causes evaporation; thus, the liquid–vapour interface recedes into the wick. This causes a reduction in the interfacial radius of curvature at the evaporator, and a circulation is set up as discussed in Chapter 2. This will pull down the level of liquid in the condenser and will tend to result in a new static balance

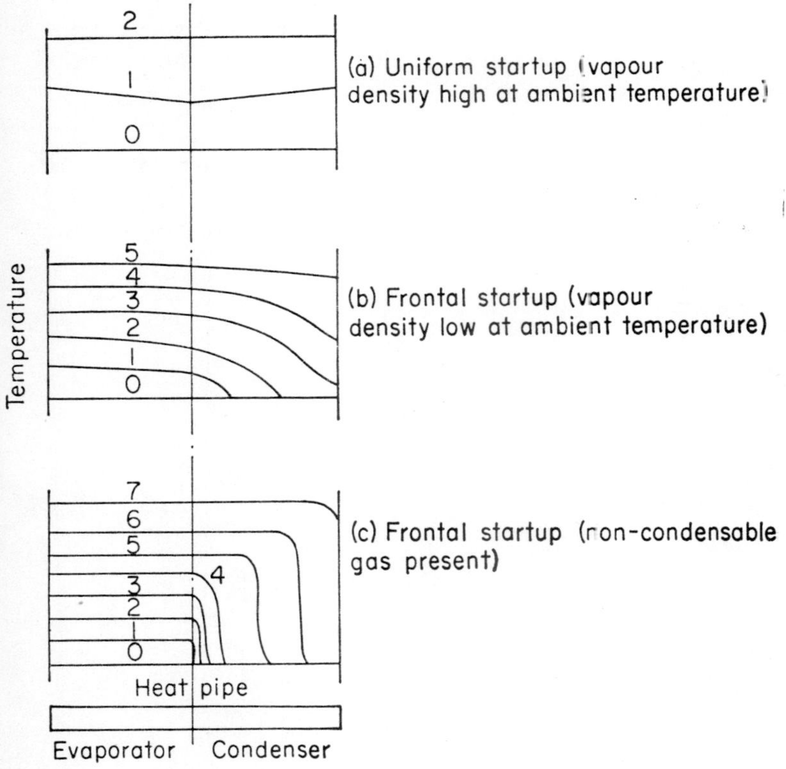

Fig. 6.1. Basic modes of Start-up.

being achieved. However, the vapour generated at the evaporator flows to the condenser, where it condenses, so that the interface there does not recede into the wick.

Cotter[15] describes three basic modes of start-up; these are illustrated in Fig. 6.1, where the number on each curve is proportional to the time from start-up. In Fig. 6.1(a), the initial vapour density is sufficiently high for an approximately uniform vapour temperature to be obtained throughout start-up. When the initial vapour density is low (Fig. 6.1(b)), "frontal start-up" is obtained; compressible flow effects can be of importance and sonic velocities can occur initially, with large axial temperature variations in the vapour. If non-condensable gases are present, the temperature variation is as shown in Fig. 6.1(c). The motion of the vapour sweeps the gases towards the condenser. This vapour motion and the increasing vapour pressure as the temperature rises combine to compress the gases into the end of the heat pipe.

Fig. 6.2 illustrates the compressible flow effects that can occur if the temperature outside the condenser is suddenly reduced. For simplicity, the case is illustrated where the

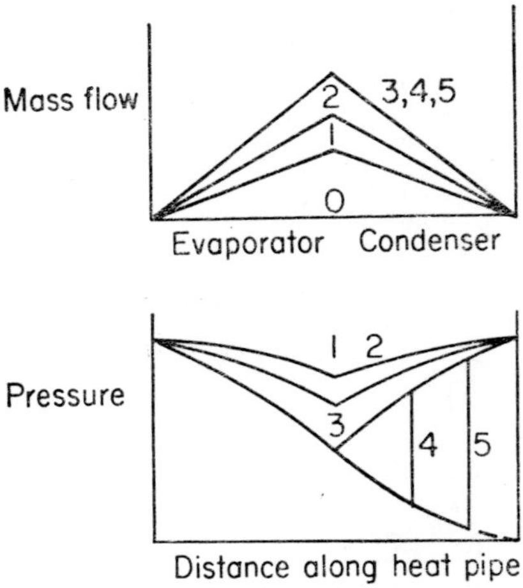

Fig. 6.2. Compressible flow effects at start-up.

pressure at the end of the evaporator remains constant. During time intervals 1, 2 and 3, the flow rate increases until the sonic velocity occurs between the condenser and evaporator. As the pressure in the condenser decreases further, supersonic flow can occur in the condenser, followed by a shock wave associated with a step change in pressure as illustrated at time intervals 4 and 5.

6.2 FLUID SOLIDIFICATION

Where, prior to start-up, the fluid is partly or entirely frozen or solidified, difficulty can be encountered in starting a heat pipe. Fig. 6.3 shows[S8] the variation in axial temperature distribution along a heat pipe of stainless steel containing water as the working fluid. Initially, the condenser is frozen, and it remains so after 11 minutes despite the evaporator temperature having risen in that time to 200°C.

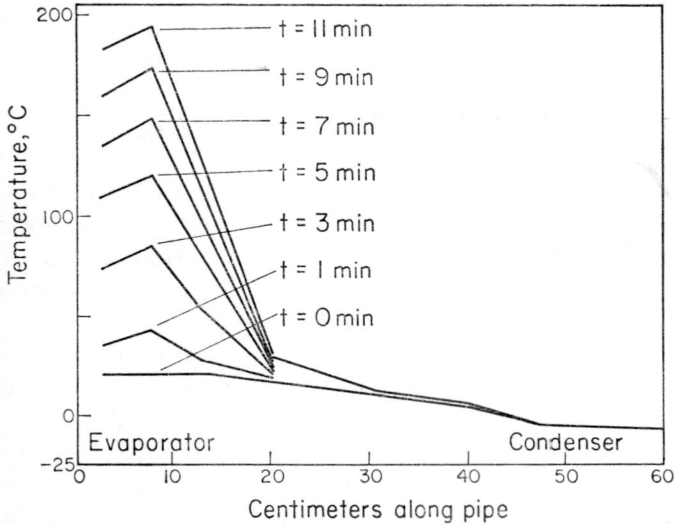

Fig. 6.3. Temperature profiles along a frozen heat pipe at different time intervals after start-up. (*Shlosinger*)

Shlosinger[8] has examined the possibility of using additives to depress the freezing point of water. Alcohols and ethylene glycol, the commonly used engine antifreezes, have a high vapour pressure; thus, they would form the greater part of the vapour mixture in the heat pipe, reducing the efficiency of the heat and mass transfer processes. This also reduces their efficiency as antifreezes. To reduce this fractionation, a search was made for a fluid which is azeotropic with water. Propanol has a vapour pressure curve almost identical to water above 0°C; in addition, it has a freezing point of −127°C. Unfortunately, to reduce the freezing point of water to −40°C, for example, it is necessary for the mixture to contain about 85 per cent by weight of propanol.

Shlosinger[7, 8] has suggested the use of a secondary heat pipe within the main heat pipe, which would have as its working fluid a fluid capable of working at a sufficiently low temperature. It would transfer heat from the evaporator to the condenser wick. This, of course, is applicable only where the fluid of the secondary heat pipe would give less efficient performance than the primary fluid under normal operating conditions.

6.3 CONSTANT TEMPERATURE HEAT PIPE

The heat pipe can be adapted readily to applications where control of heat transfer or temperature is required (see Fig. 1.7). One method uses inert gas in the vapour. As the vapour moves along to the condenser, the inert gas is swept before it, accumulating at the end of the condenser, where it blankets off heat transfer. If, for example, the source temperature decreases, the vapour temperature in the heat pipe will also tend to drop. This will cause a fall in the vapour pressure which will cause the non-condensable gas to expand, reducing the rate of heat transfer from the condenser. Fig. 6.4 illustrates the arrangement; with a suitably sized reservoir of inert gas, an almost constant vapour temperature can be maintained for large variations in axial heat transfer rate. Units have operated successfully with as much as an eight-fold variation in axial heat flux.

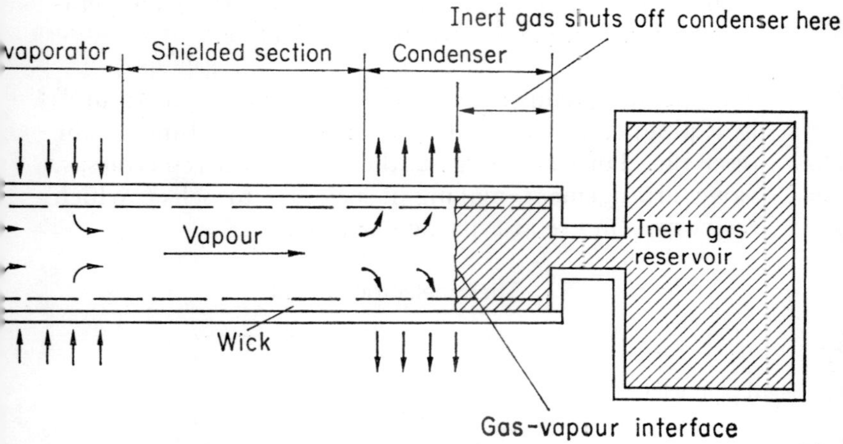

Fig. 6.4. Heat pipe with inert gas reservoir

Maintaining the vapour temperature constant has two main advantages. It tends to stabilize the source temperature, which can be advantageous, for example, in the case of electronic equipment. Also, it tends to maintain the surface at the condenser at constant temperature, which can be of advantage for example in heating temperature-sensitive fluids.

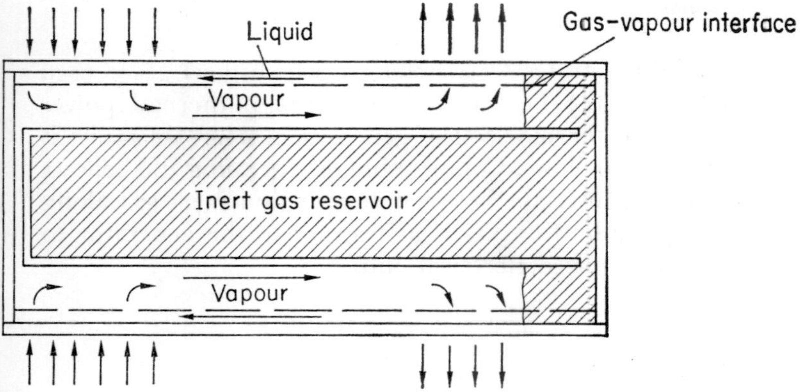

Fig. 6.5. Heat pipe with inert gas reservoir at vapour temperature

Ideally, the reservoir should be thermally coupled to the heat pipe so that the reservoir itself is maintained at approximately constant temperature, otherwise the gas blanket will expand or contract as the reservoir temperature varies. Fig. 6.5 illustrates this principle.

The operating temperature of these "constant temperature" heat pipes can be altered by changing the volume of the reservoir. Turner[(2)] suggests the use of two reservoirs, a valve arrangement making it possible to select which reservoir is in use at a particular time.

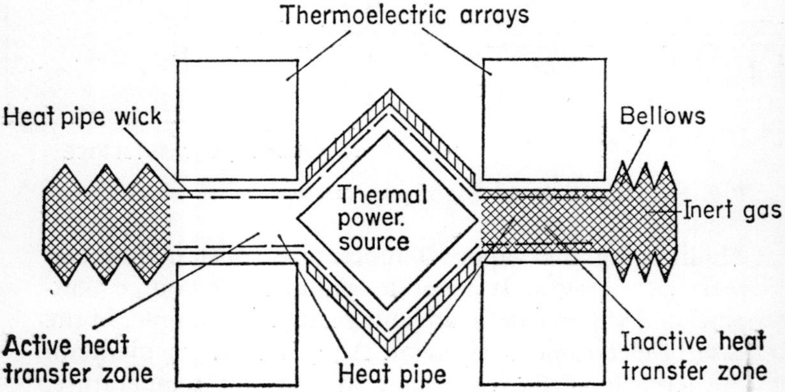

Fig. 6.6. Thermoelectric Generator with heat pipe control (Mechanically operated bellows move gas-vapour interface.) (Feldman)

Fig. 6.6 illustrates[(F6)] a system in which inert gas is used to control the heat transferred from a thermal power source to thermoelectric generators. The reservoir is formed by bellows. Compressing or extending the bellows moves the interface between vapour and inert gas along the pipe, adjusting the heat transfer to the generator to give the required electrical output.

6.4 BINARY MIXTURES

The use of binary mixtures has been examined as a means of controlling the evaporator and condenser at different temperatures. Tien has examined the performance of a heat pipe containing mixtures of water and ethanol. Fig. 6.7 shows the temperature variation obtained along the wall of the pipe when transferring axially 100 *W*. With water alone, the temperature difference between the walls in the two sections is approximately 10 *K* and with pure ethanol 2 *K*; these temperature differences occur mainly radially rather than axially. With a mixture of water (56 per cent by weight) and ethanol (44 per cent by weight), for example, the temperature difference is about 22 *K*.

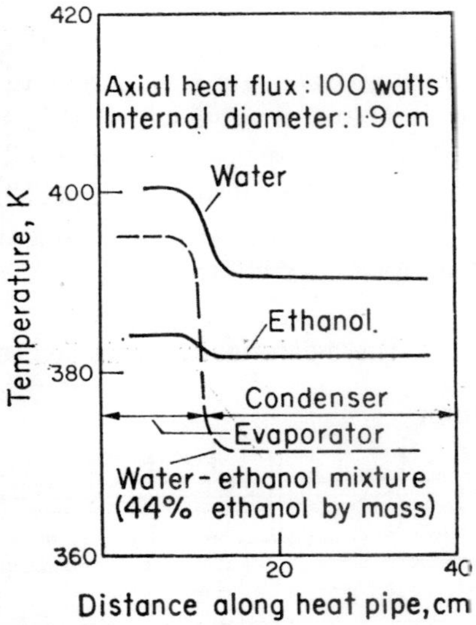

Fig. 6.7. Temperature variation along wall of heat pipe using binary mixture. (*Tien*)

6.5 THERMAL SWITCH

Another form of heat pipe(S8) can be used to control either the source or sink temperature within certain limits.

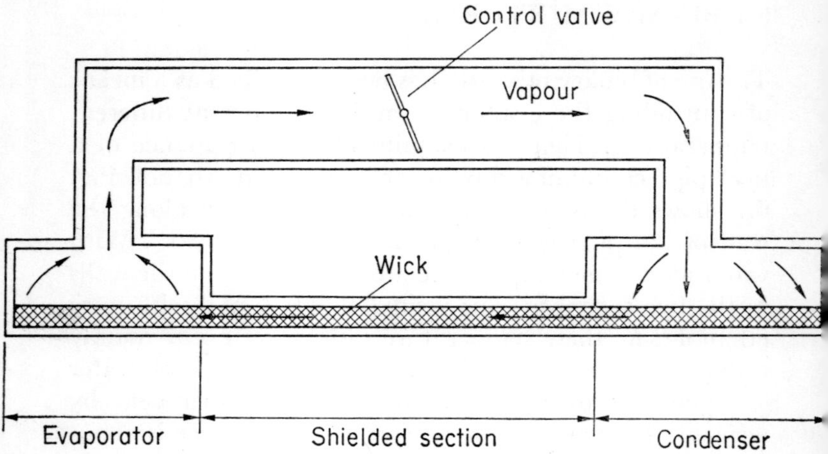

Fig. 6.8. A thermal switch

The arrangement is shown in Fig. 6.8. The liquid and vapour flow through separate tubes, and in the vapour flow tube there is a control valve which adjusts the flow to maintain the required temperature. Fig. 6.9 illustrates the use of this principle in a flat plate heat pipe. Shlosinger has carried out experiments on heat pipes using this principle for space-suit temperature control.

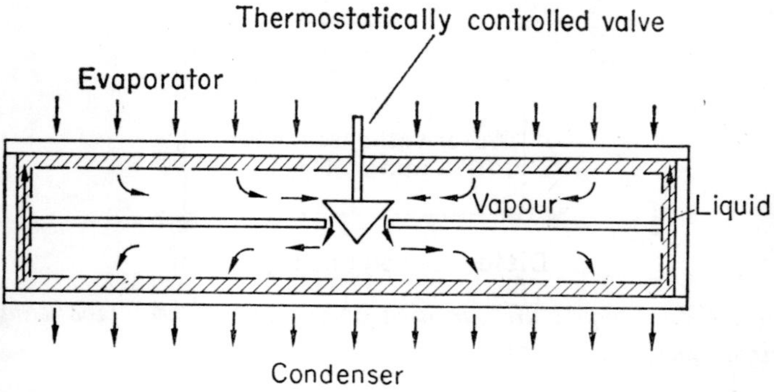

Fig. 6.9. A thermal switch: plate heat pipe

7. Manufacture and Life

7.1 THE CONTAINER

Basic requirements of the container are that it should be leakproof over the operating pressure range and have a satisfactory life. It can be made[C7] by a number of standard manufacturing techniques, such as drilling a rod in the case of a cylindrical heat pipe, casting, seam-welding from sheet, and chemical vapour deposition on a mandrel which is later dissolved. The method of manufacture depends on the selected material and the heat pipe shape.

The container material must be compatible with the working fluid, so that corrosion[B10] and evolution of gas[S6] do not occur. Corrosion can be caused in several ways. Intermetallic compounds may be formed, resulting in deterioration of the container wall and possible leakage from the pipe. The wall material may be soluble in the working fluid. It may be dissolved in the condenser and transferred in the fluid to the evaporator, where it is either deposited on the wall or the capillaries. Where it is deposited on the wall, it provides a resistance to heat transfer, and a hot spot can occur; if it is deposited in the capillary passages, it will restrict flow, reducing the performance capability of the unit and perhaps causing burnout of the pipe.

In a similar corrosion mechanism,[B10] oxygen in the wall at the cool end is extracted by the working fluid, and is carried to the hot zone, where it accumulates. Thus, while the average oxygen content of the metal may be too small to produce corrosion, this gradual accumulation may result in concentrations of oxygen so high that rapid wall attack can occur. Busse[10] has discussed this mechanism in some detail.

The material and fluid used must be such that they do not react to produce gases. Schwartz found that the performance of a heat pipe of stainless steel with water as the fluid varied with time, as shown in Fig. 3.2. For the first 31 days there was a gradual deterioration of performance, primarily in the condensing section. After 5 months, gases in the container were analysed by mass spectrometer, and were found to contain about 99 per cent hydrogen. This was attributed to the following reaction occurring in the pipe.

$$Fe + 2H_2O = Fe(OH)_2 + H_2$$

The temperature range of operation was from 30 to 160°C, which is well below the temperatures at which the water itself would decompose. The fall-off in temperature of the tube wall in the condensing section is due to the blanketing effect of the hydrogen on the heat transfer processes; the gases will be swept along by the vapour into the condenser. The generation of hydrogen by this mechanism can be avoided by choosing a material below hydrogen in the electrochemical series (e.g. copper). If a metal above hydrogen in the series is selected, it may be used with a non-reacting fluid. Stainless steel and ammonia were found by Schwartz to be a satisfactory combination.

Owing to the sensitivity of heat pipe performance on very small quantities of gas it is necessary to out-gas the container thoroughly before adding the working fluid and sealing the container. As out-gassing is easier at high temperature, the pipe material and the pipe design must be such as to withstand temperatures which may be considerably in excess of the operating temperature.

Care is required in the production of the container, as extremely small qantities of metal from the tools and jigs used in manufacture can become embedded in the container, causing intermetallic corrosion and a shortening of the heat pipe life.

It is probable that composite walls will be used in the future, including, for example, metal walls with refractory linings to prevent corrosion. In this way, strength and corrosion resistance can be provided.

7.2 THE WICK

The wick must be compatible with the fluid and with the material of the wall.

Many wick geometries are used, as discussed in Chapter 5. The design and assembly procedure must(K2) ensure that the interface between the phases, particularly in the evaporator, cannot take up a location where the capillary radius is greater than the effective pore size of the wick. This is illustrated by the diagrammatic view of the end of the heat pipe at the evaporator in Fig. 7.1; the effective capillary driving force in this example is determined by the end clearance. Similarly, where it is necessary to join two wicks together, care is needed.

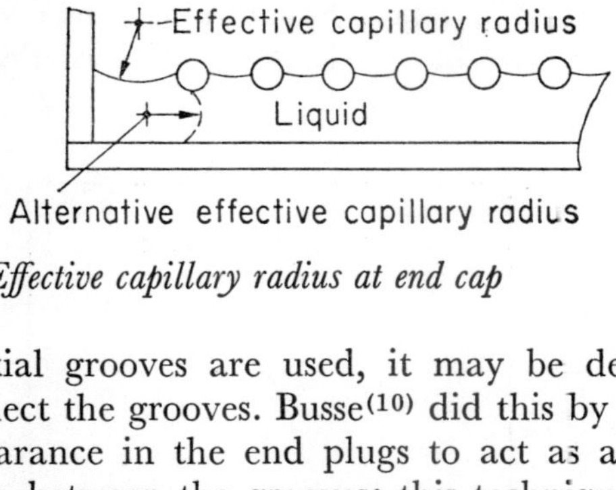

Fig. 7.1. Effective capillary radius at end cap

Where axial grooves are used, it may be desirable to inter-connect the grooves. Busse(10) did this by providing radial clearance in the end plugs to act as a capillary connection between the grooves; this technique was not used in earlier(B9) studies.

Good contact is necessary between the wick and the wall in the condenser and evaporator to keep to a minimum the temperature differences between the vapour and the tube wall. Wicks may be held in place using spiral wire springs; these are wound tight before being placed in the pipe, and they are then allowed to open, forcing the wick against the wall. A spot-welded screen tube was hammered(B10) against the pipe wall using a loose W-cylinder energized by vibrating the heat pipe in a round hammer machine. A tapered mandrel can also be used to press the wick against the wall.

The possibility of bonding the wick metallically to the pipe wall has also been examined. Katzoff[1] describes a procedure in which a coating of indium or tin is electroplated on both the screen and the wall; the two are pressed firmly together and heated in an oven at 200°C (indium) or 275°C (tin).

Diffusion bonding is another technique[K1] which may be considered for attaching the screen to the wall. With a stainless steel wick and pipe, a bond was achieved by pressing the two together with a pressure of about 1 bar and heating at 1100°C for 2 hours under vacuum. This technique is advantageous if the pipe must subsequently be subjected to high temperature in out-gassing or in welding during assembly. The wick can be pressed to the wall by using an inflatable silicone-rubber bladder at low temperature, or a bladder of stainless steel foil at high temperature. Bonding techniques are also described by Shlosinger.[8]

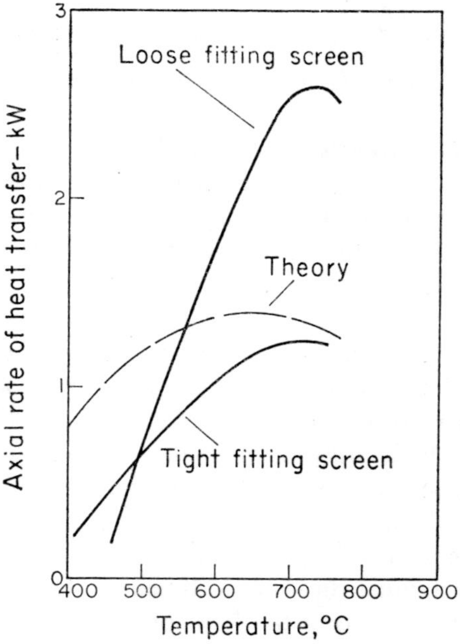

Fig. 7.2. Influence of screen fit on heat pipe performance

In considering methods of attachment it is worth noting that, using a composite wick consisting of channels covered by a screen, Kemme[3] found that a loosely fitted screen gave axial heat fluxes over twice those obtained with well-fitted screens. This is shown in Fig. 7.2. The better performance with the loosely fitting screen is due presumably to a smaller resistance to flow of the liquid. The predicted curve in Fig. 7.2 corresponds to the case of a well-fitted screen.

7.3 THE FLUID

The need for selecting a fluid and material that are compatible, to avoid corrosion and gas generation, have been discussed in the previous sections.

It is necessary to use pure fluids. If impure fluids are used, impurities more volatile than the working fluid will go to the condenser, blanketing the heat transfer surface and reducing heat transfer. The less volatile impurities will collect in the evaporator wick and impair circulation and heat transfer. Heat pipes are often filled by direct vacuum distillation in order to reduce impurities in the working fluid. Schwartz used water, distilled several times until its electric resistivity measured 10^6 ohms. Care must also be taken to ensure that the working fluid does not contain dissolved materials which would be precipitated from solution.

Sufficient fluid should be added to the container to ensure that under the operating conditions the wick is completely saturated, with sufficient surplus to flood, or ensure zero radius of curvature, over as short a length of heat pipe as possible (see Fig. 2.1).

Busse[10] precleaned lithium by melting it 2 or 3 times in a vacuum of 10^{-4} to 10^{-5} torr through a nickel funnel with a 1 mm opening; the lithium was pushed by a nickel piston. This procedure removed lithium oxide.

7.4 FILLING AND SEALING

Fig. 7.3 illustrates the apparatus used by Schwartz for filling a stainless steel heat pipe with water. Distilled water is placed in the vial which is sealed before being connected to the remainder of the apparatus. The heat pipe container and connecting pipe work are then evacuated for a number of hours, during which time the metal surfaces are heated to aid out-gassing. Valve *M* is then closed, and the seal of the water cylinder broken (see Reference S6). The vial of water is heated gently to evaporate the water, which then condenses in the container cooled with liquid nitrogen. When sufficient liquid has transferred to the heat pipe, valve *N* is closed and the unit is ready for permanent sealing.

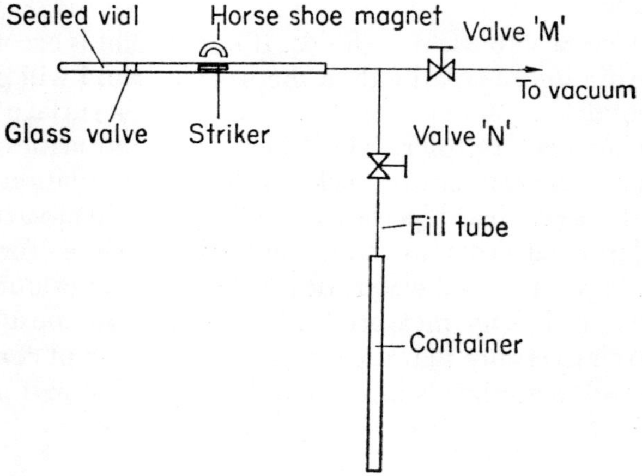

Fig. 7.3. Rig for filling container with working fluid. (*Schwartz*)

In filling heat pipes with liquid metals, Busse[9] has used the following steps:

1. The container and wick are out-gassed at high vacuum (10^{-5} torr) for 30 hours at 2000°C. Where cavities occurred, these were ground open to facilitate out-gassing.
2. After exposure to air, the container is further out-gassed for 10 minutes at around 1000°C or higher.

3. Introduction, under 99·995 per cent pure argon, of the precleaned fluid either as liquid or solid.
4. The heat pipe is out-gassed for some minutes under operating conditions at a temperature slightly below normal operating temperatures.
5. The heat pipe is then sealed.

Fig. 7.4 shows the arrangement of end cap and sealing plug used by Busse.[8] Before the addition of the working fluid to the container, the end cap is welded in place. The fluid is subsequently added by distillation through the hole in the plug. The unit is then sealed by melting the plug by electron beam, the plug flowing over the end cap to seal the hole.

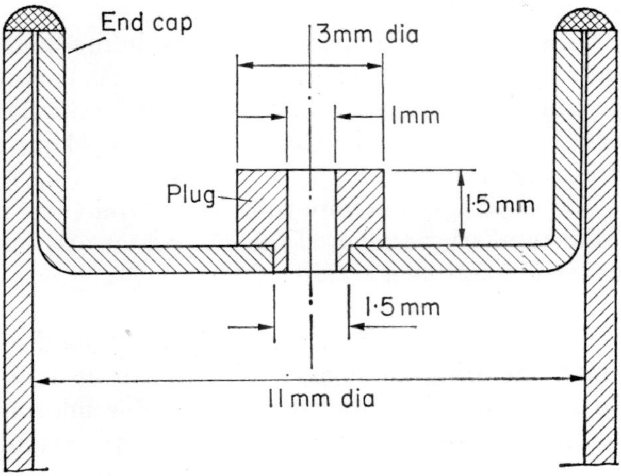

Fig. 7.4. End cap and plug. (*Busse et al.*)

With liquid metals, the metal can be distilled into the heat pipe using essentially the system discussed and shown in Fig. 7.3. The required amount of metal is placed in a distillation pot and heated until a temperature rise of the pot indicates that the distillation process is complete. The connection tube is then "pinched off" and the unit sealed. Kemme[3] subsequently heated a potassium-filled heat pipe for 60 hours at 800°C to allow the fluid to wet the capillary structure.

Busse (in Reference 10) outlines a somewhat different filling procedure.

7.5 LIFE

The life of a heat pipe is understood to mean that period of time that the heat pipe operates at its design capability. As discussed in the previous sections, life is curtailed by the following:

1. Corrosion due to impurities or incompatibility of the working fluid and container. This can result in a container leak or deterioration in performance due to corrosion deposits clogging the capillaries and reducing circulation, or to the corrosion impeding heat transfer through the wall.
2. Gases collecting in the vapour space in the condenser and reducing heat transfer. These gases may come from the container wall if the container is not initially outgassed, or they may be due to reactions taking place within the heat pipe.
3. High boiling-point impurities in the liquid may move to the capillary passages of the evaporator and impede circulation and heat transfer there.

Providing the above effects are avoided by good design and manufacturing methods, long lives can be achieved. Table 7.1 summarizes some of the available information related to heat pipe life-times; there are many examples of lives in excess of 15,000 hours being achieved.

Table 7.1. Some heat pipe life tests

FLUID	CONTAINER	REMARKS	SOURCE
Water	S.S.	Rapid deterioration in condenser performance over first 31 days due to hydrogen generation	Schwartz
Dowtherm E	Not known S.S. Wick	15,000 hours at 300°C (?). No failure. Similarly with a copper wick	Basiulis
Pb	Nb	3 hours at 1600°C. Intercrystalline corrosion	Busse[8]
1Zr/Ph	Nb	10 hours at 1600°C. Dry-out	
K	Ni	16,000 hours at 600°C	
K	Berylium	1200 hours at 750°C. No failure but recrystallization and grain growth had occurred	
Silver	Tungsten	1000 hours at 1900°C. Negligible mass transfer	Grover[13]
Indium } Silver }	Rhenium Tungsten Wick	1000 hours at 2000°C. Considerable mass transfer Considerable mass transfer at 300 hours	
Li } Li-13Ca }	Deoxidized Nb-1Zr	Both 1000 hours at 1500°C. Zr loss	Busse[10]
Li	SGS-TA	1000 hours at 1500°C. No Zr loss	

8. Commercial Heat Pipes

A wide variety of heat pipes is now available commercially. The names and addresses of a number of manufacturers are given in Appendix II; at the present time (1971), as far as the writer is aware, heat pipes are not being produced commercially in the U.K. Typical examples of available heat pipes are given in Table 8.1. Sizes range from 0·13 to 3·8 cm outer diameter and up to 6 m long (Noren Products). The range of container materials includes copper, nickel, stainless steel, aluminium, Pyrex and ceramic. The temperature range covered is from about −170 to 1000°C and above. The fluids include water, ammonia, alcohols, sodium and lithium.

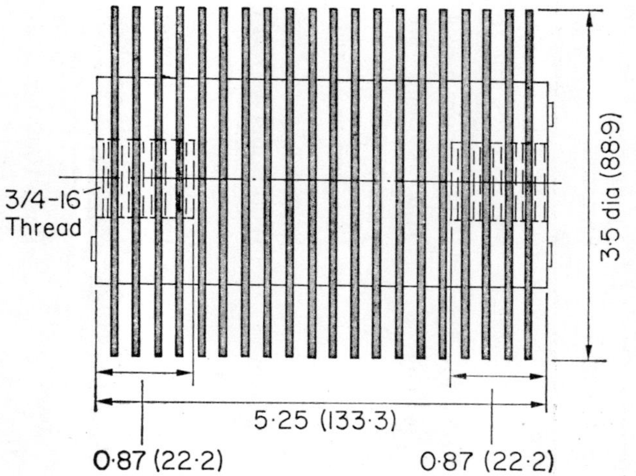

Fig. 8.1. Heat pipe-heat sink: RCA Unit No J 15121
Dimensions are in inches unless otherwise stated. Dimensions in parentheses are in millimeters and are derived from the basic inch dimensions (1 inch = 25·4 mm)

Heat pipes from stock usually have plain ends when they are of the simple tubular type, but heat pipes with lugs, screw threads and other end shapes are available. Heat pipes that can be bent using a standard copper tube

Table 8.1. Typical commercial tubular heat pipes

NUMBER	FLUID	CONTAINER	OUTSIDE DIAMETER cm	LENGTH m	TEMPERATURE °C	TEMPERATURE DIFFERENCE K	CAPACITY Watts	RADIAL FLUX Watt/cm² EVAPORATOR	RADIAL FLUX Watt/cm² CONDENSER	ORIENTATION	REFERENCE
MG12		Glass	1·2	0·203	—40–80		50			Horizontal	Energy Conversion Systems, Inc.
WG12		Glass	1·2	0·203	20–120		35				
WC1/8		Copper	0·317	0·152	20–230		6				
WC3/4		Copper	1·905	0·457	20–230		350				
	Water	304 S.S.	0·635	0·178	43	15	40			Opposing Gravity	Thermo Electron Corporation
	Sodium	304 S.S.	1·27	0·305	700	3·6	750			Horizontal	
	Lithium	TZM		0·305	1500	5·0	4000			Arbitrary	
J15105	Sodium	300 S.S.	1·11	0·374	900		1000	30	30	Horizontal	R.C.A. Ltd
SS59		S.S.	0·15	0·203	−55–120			30	15		Noren Products
06-175-SR-D		S.S.	0·94	0·445	150–400						
		*	1·27		1000		380				
			1·27		1000		190*				
			2·54		1000		6000				

* Bendable

bender are also made; the bend causes some decrease in performance (see Table 8.1). Flexible heat pipes are also available. Noren Products, for example, produce units with a 25 cm long metal bellows in the middle of the pipe. They also produce electrically insulated heat pipes which have a 13 cm length of glass section in the middle.

Heat pipes with external finning can be obtained. Fig. 8.1 illustrates a unit produced by RCA (Unit No J15121). It can be used as a heat sink with two heat sources screwed into the ends, or as a means of coupling together a hot source and cold sink, while at the same time facilitating heat transfer from the system. The operating temperature is 150°C maximum; 150 W and 1000 W can be dissipated with natural convection and with forced air cooling respectively.

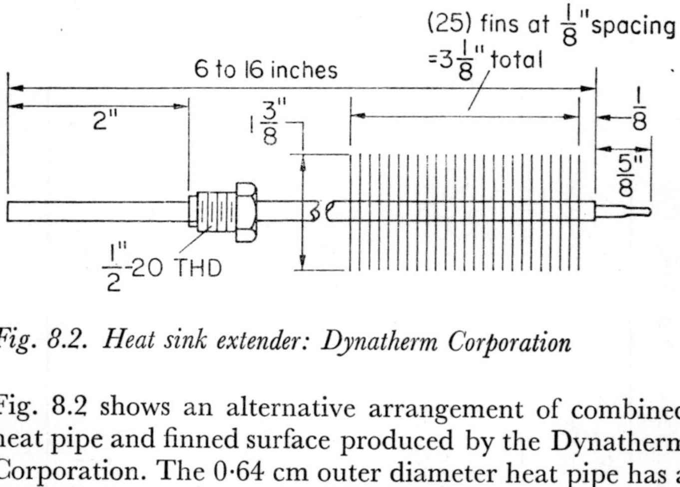

Fig. 8.2. Heat sink extender: Dynatherm Corporation

Fig. 8.2 shows an alternative arrangement of combined heat pipe and finned surface produced by the Dynatherm Corporation. The 0·64 cm outer diameter heat pipe has a limit of capacity of 35 W. The evaporator section of the heat pipe screws into the body of the unit to be cooled. Fig. 8.3 shows the performance characteristics of the unit.

Fig. 8.4 shows a unit in which a heat pipe acts as a "coupler" between a heat source and a heat sink, while in addition acting as a thermal transformer by increasing the effective sink surface area (RCA Unit No J15120). It

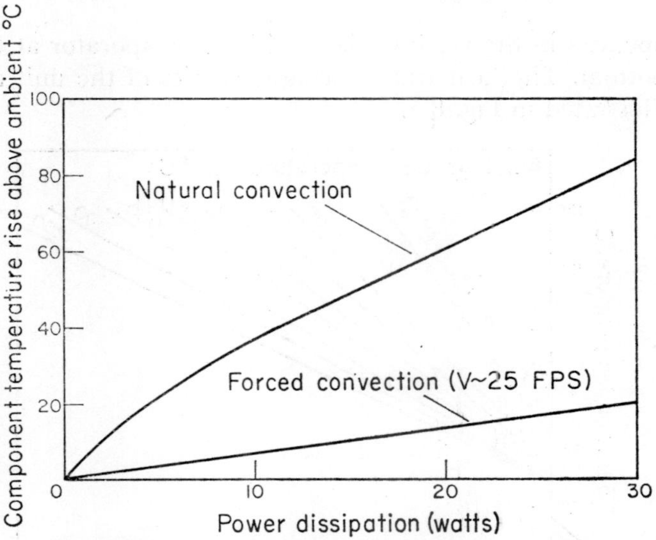

Fig. 8.3. Performance characteristics of Dynatherm heat sink extender

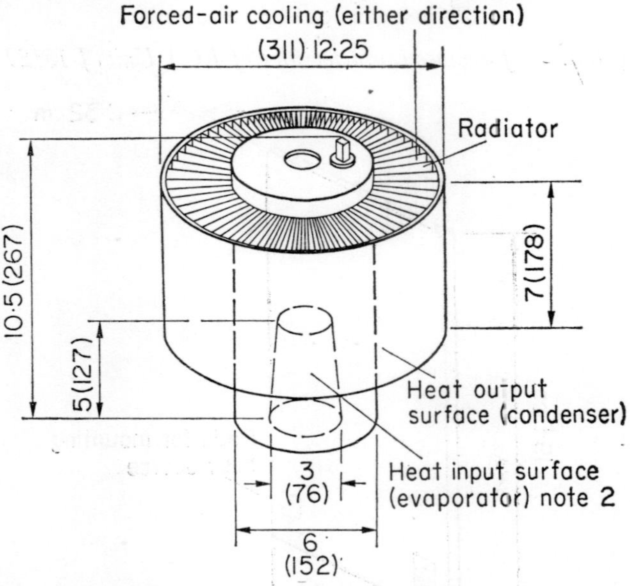

Fig. 8.4. Heat pipe-heat sink: RCA Unit No J 15120

Note 1: Dimensions are in inches unless otherwise stated. Dimensions in parentheses are in millimeters and are derived from the basic inch dimensions (1 inch = 25·4 mm)
Note 2: The taper on the heat input surface is 1·5° unless otherwise requested.

operates in the vertical plane with the evaporator at the bottom. The heat transfer characteristics of the unit are illustrated in Fig. 8.5.

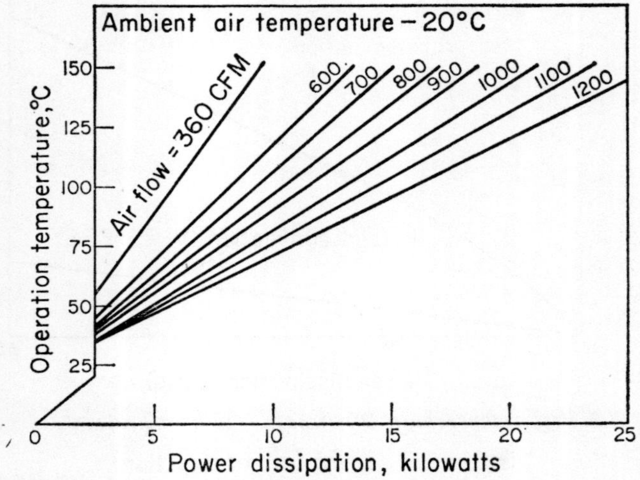

Fig. 8.5. Performance characteristics of RCA Unit J 15120

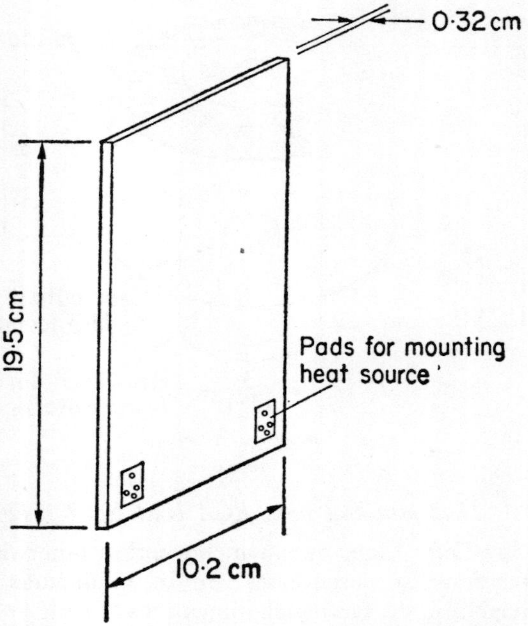

Fig. 8.6. Flat plate heat pipe: Noren Products Unit No FC408

Flat heat pipes are available with dimensions up to 61 cm by 66 cm on which transistors or other hot devices can be bolted as indicated in Fig. 8.6 (Noren Products). The performance of unit No FC408 is given in Fig. 8.7. Two

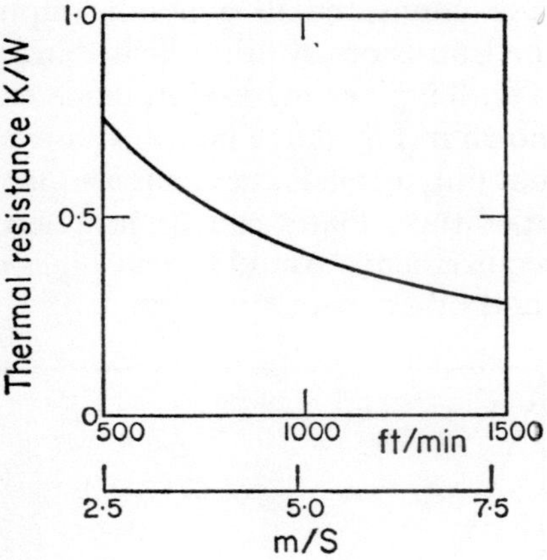

Fig. 8.7. Performance characteristics of flat heat pipe: Noren Product Unit No FC408

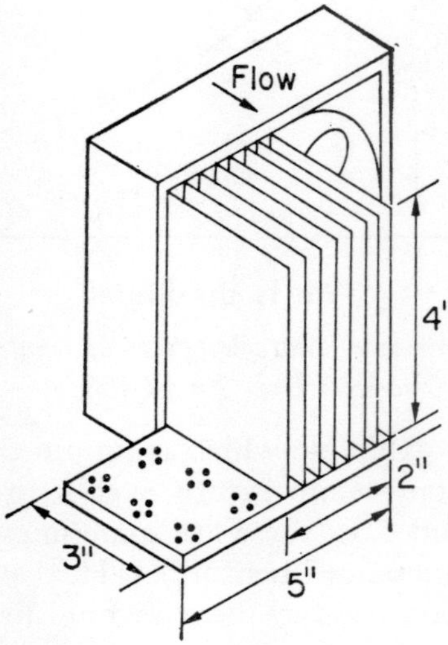

Fig. 8.8. Multi-pin heat pipe assembly: Noren Products MF Series

different types of pad for attaching the heat source are available: holes can be made on the plate in the factory, or solid pads can be supplied to be drilled as required by the user. Multi-fin arrangements are also available from Noren Products as illustrated in Fig. 8.8; the fins, which are flat heat pipes, are connected to a hollow input pad by a manifold. The transistors or other heat sources are bolted to the pad. Fig. 8.9 gives the performance characteristics of the unit shown in Fig. 8.8. The transistors or other heat sources are not only cooled; they will also tend to be at a uniform temperature. Plates can be supplied bent to the customer's requirements. Annular heat pipes are made by Dynatherm and other manufacturers.

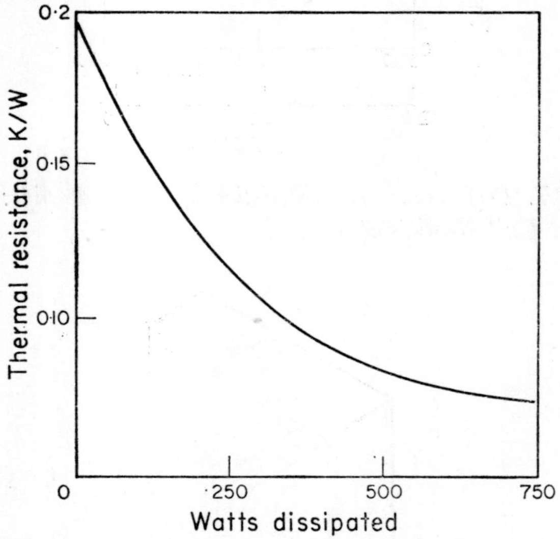

Fig. 8.9. Performance characteristics of multi-fin heat pipe assembly: Noren Products Unit No FC408

Heat pipes are available which maintain themselves at a relatively constant temperature over a range of output (see Figs 6.4 and 6.5); these are sometimes referred to as "variable conductance" heat pipes. Heat pipe diodes are heat pipes which conduct heat in one direction. Noren Products claim that the ratio of effective pipe thermal conductivities in their diodes in opposite directions is 1000 : 1.

Q-dot Corporation of Santa Fe, New Mexico, market a heat exchanger for use in air-conditioning systems; it is shown diagrammatically in Fig. 8.10. The heat is transferred between the two air flow streams by an array of heat pipes.

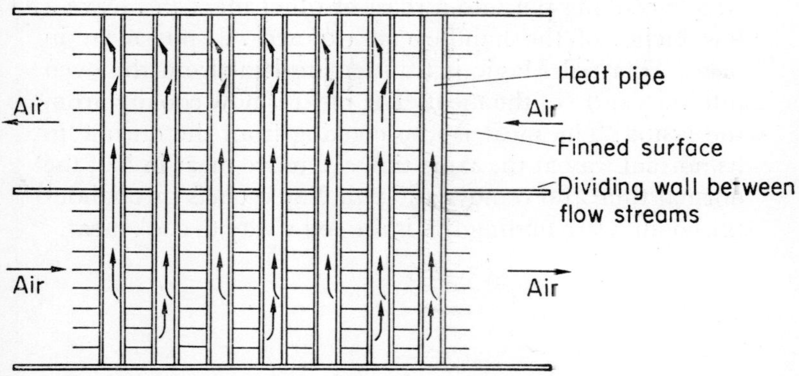

Fig. 8.10. Air to air heat exchanger for air conditioning system: Q-dot Corporation

Scott Engineering Sciences of Pompano Beach, Florida, produce a "Heat Pipe Education System". The heat source in the system, which is mounted on a panel, is a 200 W electrically heated oven, and the sink is a water-jacket. The "master" heat pipe, which is made of brass, is 1·6 cm diameter and 46 cm long. It has removable end caps so that the wick can be changed; it is instrumented with seven thermocouples and a pressure gauge. A needle-type bellows-valve allows change of the working fluid and the initial pressure in the pipe. Three demonstration heat pipes are also supplied, one with a glass wall and stainless steel wick with methyl alcohol as the fluid, another made of copper with the same working fluid, and a third of copper with water as the working fluid.

Noren Products make a system designed and constructed to demonstrate the operating principles of the "variable conductance" heat pipe. The heat pipe is of Pyrex glass, so that the interior is visible. A half-wick is used in the condenser, enabling the vapour passage to be seen. A coloured inert gas is used to distinguish it from the condensable working fluid.

Last, but not least, there is a heat pipe product advertised by Energy Conversions Systems, Inc.: "Insert the Thermal Magic cooking pin into a roast or other meat . . . leave a few inches of the handle end exposed . . . put meat in oven. Thermal Magic is transferring heat from the oven into the centre of the meat. The meat is now cooking from the INSIDE! The meat is also cooking from the outside in its normal way at the same time. Remove meat in half the normal time and remove the pin. Noted chefs throughout the country are finding this is the ONLY way to cook meat."

Bibliography

1. ALLINGHAM, W. D., MCENTIRE, J. A. Determination of boiling film co-efficient for a heated horizontal tube in a water-saturated wick material. *J. Heat Transfer*, Feb. 1961, **83** (1), 71–75.
2. AMERICAN SOCIETY OF HEATING, REFRIGERATING AND AIR CONDITIONING ENGINEERS. *Handbook of Fundamentals*, New York, N.Y.: ASHRAE, 1967, 544 pp.
3. ANAND, D. K. On the performance of a heat pipe. *J. Spacecraft and Rockets*, May 1966, **3**(5), 763–765.
4. ANAND, D. K. Heat pipe application to a gravity-gradient satellite (Explorer XXXVI). *ASME—Aviation and Space—Progress and Prospects*, Annual Aviation and Space Conference, Beverley Hills, Calif., 16–19 June 1968, pp. 634–638.
5. ANAND D. K., DYBBS, A. Z., JENKINS, R. E. Effects of condenser parameters on heat-pipe optimization. *J. Spacecraft and Rockets*, May 1967, **4**(5), 695–696.
6. ANAND, D. K. *et al.* Heat pipe application for spacecraft thermal control. *JHU Tech Memo TG922.* Contract Now 62 0604, Silver Springs, Md: John Hopkins Univ., App. Physics Lab. Aug 1967. AD 662 241.

1. BÄHR, A., BURCK, E., HUFSCMIDT, W. Ä. Liquid–vapour interaction and evaporation in heat pipes. Inter. Conf. on Thermionic Elec. Power Generation, 2nd, Stresa, Italy, 27–31 May 1968. Proceedings. Conference sponsored by European Nuclear Energy Agency. Luxembourg, EURATOM Centre for Information and Documentation (EUR No 4210 e, f) 1969, pp. 543–556.
2. BAINTON, K. F. Experimental heat pipes. AERE–M1610. Harwell, Berks: Atomic Energy Research Establishment, Applied Physics Division, 1965.
3. BASIULIS, A., DIXON, J. C. Heat pipe design for electron tube cooling. ASME Paper 69-HT-25. New York, N.Y.: American Society of Mechanical Engineers, 1969.
4. BOHDANSKY, J., SCHINS, H. E. J. Heat transfer of a heat pipe operating at emitter temperatures. IEE—1st Int. Conf. on Thermionic Electrical Power Generation, London, 1965, pp. 1–5. London: Institution of Electrical Engineers, 1965.
5. BOHDANSKY, J., STRUB, H., VAN ANDEL, E. Heat transfer measurements using a sodium heat-pipe working at low

vapour pressure. IEEE—Thermionic Conversion Specialist Conference, Houston, 3–4 Nov. 1966, pp. 144–148.

6. BUSSE, C. A. Pressure drop in the vapour phase of long heat pipes. IEEE-Thermionic Conversion Specialist Conference. Palo Alto, Calif.; 30 Oct.–1 Nov 1967, pp. 391–398.
7. BUSSE, C. A. Heat pipe research in Europe. Int. Conf. on Thermionic Electrical Power Generation, 2nd, Stresa, Italy, 27–31 May 1968. Proceedings. Conference sponsored by European Nuclear Energy Agency. Luxembourg, EURATOM Centre for Information and Documentation (EUR No 4210 f, e), 1969, pp. 461–475.
8. BUSSE, C. A., CARON, R., GEIGER, F., PÖTZSCHKE, M. Performance studies on heat pipes. IEEE-1st Int. Conf. on Thermionic Electrical Power Generation, London, 1965.
9. BUSSE, C. A., GEIGER, F., PÖTZSCHKE, M., QUATAERT, D. Heat pipe life tests at 1600°C and 1000°C. IEEE-Thermionic Conversion Specialist Conference, Houston, 3–4 Nov. 1966, pp. 149–158.
10. BUSSE, C. A., GEIGER, F., STRUB, H. *et al.* High temperature lithium heat pipes. Int. Conf. on Thermionic Electrical Power Generation, 2nd Stresa, Italy, May 1968. Proceedings. Sponsored by European Nuclear Energy Agency. Luxembourg, EURATOM Centre for Information and Documentation (EUR No 4210 f, e), 1969, pp. 495–506.

1. CALIMBAS, A. T., HULETT, R. H. An avionic heat pipe. ASME Paper 69–HT–16. New York, N.Y.: American Society of Mechanical Engineers, 1969.
2. CARLSON, G. Heat pipe radiator design. Space Power Note No 214, USAEC Rep UCID-15165, Contract W-7405-eng-48. Livermore, Calif.: Lawrence Radiation Laboratory, Calif. Univ., June 1967.
3. CARNESALE, A., COSGROVE, J. H., FERREL, J. K. Operating limits of the heat pipe. S.C.-M-66-623. Contract AT (29–1)-789. Proceedings of Joint Atomic Energy Commission/Sandia Laboratories Heat Pipe Conference held at Albuquerque; New Mexico, June 1966, pp. 27–44. Albuquerque, New Mexico, Sandia Corp., Oct 1966.

4. CASWELL, B. F., BALZHISER, R. E. The critical heat flux for boiling liquid water systems. *Chem. Eng. Progr.*, Symp. Series on Heat Transfer, 1966, **62**(64), 41–46.
5. CHATO, J. C., STRECKERT, J. H. Performance of a wick-limited heat pipe ASME paper 69-HT-20. New York,

N.Y.: American Society of Mechanical Engineers, Aug. 1969.

6. CHEMICAL ENGINEERING. New heat-exchanger tubes make their commercial debut. *Chemical Engineering*, March 1970, **77**(5), 60–62.
7. CHEUNG, H. A critical review of heat pipe theory and applications. USAEC UCRL 50453, UC-4 (Chemistry) Livermore, Calif: Lawrence Radiation Lab., Univ. of Calif., 1968.
8. CHISHOLM, D. A survey on the condensation of superheated steam. Conference on the condensation of superheated steam, National Engineering Laboratory, 14 and 15 March 1961, pp. 7–38. Edinburgh: Her Majesty's Stationery Office, 1962.
9. CONWAY, E. C., KELLEY, M. J. A continuous heat pipe for spacecraft thermal control. *Aviation and Space: Progress and Prospects;* Proc. of the Annual Aviation and space Conf., Beverley Hills, Calif., 16–19 June 1968, pp. 655–658. New York, N.Y.: American Society of Mechanical Engineers, 1968.
10. COSGROVE, J. H. Engineering design of the heat pipe. Ph.D. Thesis, Raleigh, Carolina: Univ. of North Carolina, 1966. Order No 67-11,270. Ann Arbor, Mich.: University Microfilms Inc., 1967.
11. COSGROVE, J. H., FERREL, J. K., CARNESALE, A. Operating characteristics of capillarity-limited heat pipes. *J. nucl. Energy*, 1967, **21**, 547–558.
12. COSTELLO, C. P., FREA, W. J. The noles of capillary wicking and surface deposits in the attainment of high pool, boiling burn out heat fluxes. AI Ch E JL, May 1964, **10**(3), 393–398.
13. COSTELLO, C. P., REDEKER, E. R. Boiling heat transfer and maximum heat flux for a surface with coolant supplied by capillary wicking. *Chem. Eng. Progr.*, Symp. Series, 1963, **50**(41), 104–113.
14. COTTER T. P. Theory of heat pipes. USAEC Ref. LA-3246-MS. Contract W-7405-eng-36. Los ALAMOS, New Mexico; Los Alamos Scientific Laboratory, Univ. of Calif., 1965.
15. COTTER, T. P. Heat pipe start-up dynamics. USAEC Ref. LA-DC-9026; Conf.-671045-2. Los Alamos, New Mexico: Los Alamos Scientific Laboratories, Univ. of Calif., 1967.
16. COTTER, T. P. *et al.* Status report on Theory and experiments on heat pipes at Los Alamos. USAEC Ref. LA-DC-7206: Contract W-7405-eng-36. Los Alamos, New

Mexico: Los Alamos Scientific Laboratories, Univ. of Calif., 1965.

1. DEVERALL, J. E. Effect of vibration on heat pipe performance. USAEC Ref. LA-3798: Cont. W-7405-eng-36. Los Alamos, New Mexico: Los Alamos Scientific Laboratory, Univ. of Calif., 1967.
2. DEVERALL, J. E. Total hemispherical emissivity measurements by the heat pipe method. USAEC Ref. LA-3834-MS; Contract W-7405-eng-36. Los Alamos, New Mexico: Los Alamos Scientific Laboratory, Univ. of Calif., 1968.
3. DEVERALL, J. E., KEMME, J. E. High thermal conductance devices utilizing the boiling of lithium and silver. USAEC Ref. LA-3211. Contract W-7405-eng-36. Los Alamos, New Mexico: Los Alamos Scientific Laboratory, Univ. of Calif. 1964.
4. DEVERALL, J. E., KEMME, J. E. Satellite heat pipe. USAEC Ref. LA-3278. Contract W-7405-eng-36. Los Alamos, New Mexico: Los Alamos Scientific Laboratory, Univ. of Calif, 1965.
5. DEVERALL, J. E., SALMI, E. W. Heat performance in a space environment. USAEC Ref. LA-DC-9028. Contract W-7405-eng-36. Los Alamos, New Mexico: Los Alamos Scientific Laboratory, Univ. of Calif., 1967.
6. DEVERALL, J. E., SALMI, E. W., KNAPP, R. J. Orbital heat pipe experiment. USAEC Ref. LA-3714: Contract W-7405-eng-36. Los Alamos, New Mexico: Los Alamos Scientific Laboratory, Univ. of Calif., 1967.
7. DEVERALL, J. E., SALMI, E. W., KNAPP, R. J. Heat pipe performance in a zero gravity field. *J. Spacecraft and Rockets*, Nov. 1967, **4**(11), 1556–1557.
8. DUTCHER, C. H. JR., BURKE, M. R. Heat pipes—a cool way to cool circuitry. *Electronics*, 16 Feb. 1970, **43**(4), 94–100.
9. DZAKOWIC, G. S., TANG, Y. S., ARCELLA, F. G. Experimental study of vapor velocity limit in a sodium heat pipe. ASME paper 69-HT-21. New York, N.Y.: American Society of Mechanical Engineers, 1969.

1. EASTMAN, G. Y. The heat pipe. *Scient. Am.*, May 1968, **218**(5), 38–46.
2. ELROD, W. C., *et al.* Boiling heat-transfer data at low heat flux. *J. Heat Transfer*, 1967, **89**(3), 235–243.
3. ERNST, D. M. Evaluation of theoretical heat pipe performance. IEEE-Thermionic Conversion Specialist Conference, Palo Alto, Calif., 30 Oct.–1 Nov. 1967, pp. 349–354.

1. FARRAN, R. A., STARNER, K. E. Determining wicking properties of compressible materials for heat pipe applications. *Aviation and Space: Progress and Prospects*; Proceedings of the Annual Aviation and Space Conf., Beverley Hills, Calif., 16–19 June 1968. New York, N.Y.: American Society of Mechanical Engineers, 1968.

2. FELDMAN, K. T., JR. Scott heat pipe analysis, design and experiments. Publication No. 9068, Pompano Beach, Florida: Scott, Engng. Sciences, Dec. 1968.

3. FELDMAN, K. T., WHITING, G. H. The heat pipe. *Mech. Engng*, Feb. 1967, **89**, 30–33.

4. FELDMAN, K. T., WHITING, G. H. The heat pipe and its potentialities. *Engrs Dig.*, Lond., March 1967, **28**(3), 85–96.

5. FELDMAN, K. T., WHITING, G. H. Heat pipe. USAEC. Ref. SC-TM-66-2632. Contract AT(29-1)-789. Albuquerque, New Mexico; Sandia Lab., 1967.

6. FELDMAN, K. T., WHITING, G. H. Applications of the heat pipe. *Mech. Engng*, Nov. 1968, **90**, 48–53.

7. FERREL, J. K. A study of the operating characteristics of the heat pipe. Report of progress during the ninth quarter. USAEC Ref. ORO-3411-9. Contract AT (40-1)-3411. Raleigh, North Carolina: North Carolina State University, 1967.

8. FERREL, J. K., ALLEAVITCH, J. Vaporization heat transfer in capillary wick structures. Eleventh National Heat Transfer Conference AIChE-ASME, Minneapolis, Minnesota, 3–6 Aug. 1969. New York, N.Y.: American Institute of Chemical Engineers, 1969.

9. FERREL, J. K., CARNESALE, A. A study of the operating characteristics of the heat pipe. Fifth quarterly progress report. USAEC Ref. TID-23503. Contract AT (40-1)-3411. Raleigh, North Carolina: North Carolina State University 1966.

10. FERREL, J. K., CARNESALE, A. A study of the operating characteristics of the heat pipe. Sixth quarterly progress report. USAEC, Ref. TID-23809: Contract AT (40-1)-3411. Raleigh, North Carolina: North Carolina State University, 1967.

11. FERRELL, J. K., CARNESALE, A. A study of the operating characteristics of the heat pipe. Quarterly progress report Nos. 7 and 8. USAEC Ref. TID 23937. Contract AT (40-1)-3411. Raleigh, North Carolina: North Carolina State University, 1967.

12. FERRELL, J. K., *et al.* A study of the operating characteristics

of the heat pipe Part 1. An analytical model for the prediction of operating limits of heat. USAEC Ref. ORO-3411-12 (Pts 1 and 2). Contract AT (40-1)-3411. Raleigh, North Carolina: North Carolina State University, 1969.

13. FERREL, J. K., *et al.* A study of the operating characteristics of the heat pipe Part III. Vaporization heat transfer from flooded wick covered surfaces. USAEC Ref. ORO-3411-12 (Pt 3). Contract AT (40-1)-3411. Raleigh, North Carolina: North Carolina State University, 1969.
14. FERREL, J. K., SCHOENBORN, E. M. A study of the operating characteristics of the heat pipe. Capillarity in porous media. Report on progress during the tenth quarter. USAEC Ref. ORO-3411-10: Contract AT (40-1)-3411. Raleigh, North Carolina: North Carolina State University, 1968.
15. FOX, H. W., HARE, E. F., ZISMAN, W. A. Wetting properties of organic liquids on high-energy surfaces. *J. Phys. Chem.*, 1955, **59,** 1097–1106.
16. FRANK, S., Optimisation of a grooved heat pipe. 2nd Intersociety Energy Conversion Engineering Conference, Miami Beach, Fla., 13–17 Aug. 1967, pp. 833–845. New York, N.Y.: American Society of Mechanical Engineers, 1967.
17. FRANK, S., SMITH, J. T., TAYLOR, K. M. Heat pipe design manual. MND-3288. Baltimore: Martin-Marietta Corp., 1967.

1. GALOWIN, L. S., BARKER, V. A. Heat pipe channel flow distributions. ASME Paper 69-HT-22. New York, N.Y.: American Society of Mechanical Engineers, 1969.
2. GAUGLER, R. S. Heat transfer device U.S. Patent 2, 350, 348. Application date 21 Dec. 1942. Accepted 6 June 1944.
3. GILMOUR, G. H. Nucleate boiling—a correlation. *Chem. Engng Prog.*, Oct 1958, **54**(10), 77–79.
4. GINWALA, K., BLATT, T. A., BILGER, R. W. Engineering study of vapour cycle cooling components for space vehicles. Tech. Doc. Ref. ASD-TDR-63-582; Contract AF 33 (616)-8466. Ohio: Wright-Patterson Air Force Base, Sept. 1963.
5. GOODYKOONTZ, J. H., DORSCH, R. G. Local heat transfer coefficients and static pressures for condensation of high-velocity steam within a tube. NASA TN D-3953. Cleveland, Ohio: Lewis Research Center; Washington, D.C.: National Aeronautics and Space Administration, 1967.

6. GORRING, R. L., CHURCHILL, S. W. Thermal conductivity of heterogeneous materials. *Chem. Engng Prog.*, July 1961, **57**(7), 53–59.

7. GRAY, V. H. Feasibility study of rotating boiler for high-performance Rankine cycle-power generation systems. 2nd Intersociety Energy Conversion Engineering Conference, Miami Beach, Fla., 13–17 Aug. 1967, pp. 145–149. New York, N.Y.: American Society of Mechanical Engineers, 1967.

8. GRAY, V. H. The rotating heat pipe—a wickless, hollow shaft for transferring high heat fluxes. ASME Paper 69-HT-19. New York, N.Y.: American Society of Mechanical Engineers, 1969.

9. GRAY, V. H., JOSLYN, A. W., MARTO, P. J. Boiling heat transfer coefficients interface behaviour and vapour quality in rotating boiler operating to 475 G's. NASA TN-D-4136. Contract 120-27-02-03-22. Cleveland, Ohio: Lewis Research Center; Washington D.C.: National Aeronautics and Space Administration, 1968.

10. GRIFFITH, P., WALLIS, J. D. The role of surface conditions in nucleate boiling. Third National Heat Transfer Conference ASME-AIChE, Storrs, Connecticut, 9 to 12 Aug. 1959, Reprint 106. New York, N.Y.: American Institute of Chemical Engineers, 1959.

11. GROVER, G. M., BOHDANSKY, J., BUSSE, C. A. The use of a new heat removal system in space thermionic power supplies. EUR 2229e. Ispra, Italy: European Atomic Energy Community, Euratom, Joint Nuclear Research Centre, 1965.

12. GROVER, G. M., COTTER, T. P., ERIKSON, G. F. Structures of very high thermal conductance *J. Appl. Phys.*, 1964, **35**(6), 1990–1991.

13. GROVER, G. M., KEMME, J. E., KEDDY, E. S. Advances in heat pipe technology Int. Conf. on Thermionic Electrical Power Generation, 2nd, Stresa, Italy, May 1968. Proceedings, sponsored by European Nuclear Energy Agency. Luxembourg, EURATOM Centre for Information and Documentation (EUR No 4210 e, f) 1969, pp. 477–490. Discussion, pp. 491–494.

1. HALLER, H. C., LIEBLEIN, S., LINDOW, B. G. Analysis and evaluation of a vapour-chamber fin-tube radiator for high-power Rankine cycles. NASA TN D-2836. Cleveland, Ohio: Lewis Research Center; Washington, D.C.: National Aeronautics and Space Administration, 1965.

2. HALLER, H. C., LIEBLEIN, S. Feasibility studies of space radiators using vapour chamber fins. SC-M-66-623: Contract AT (29-1)-789. Proceedings of Joint Atomic Energy Commission/Sandia Laboratories Heat Pipe Conference held at Albuquerque, New Mexico, 1 June 1966. Albuquerque, New Mexico: Sandia Corp., Oct. 1966.

3. HARKNESS, R. E. The Geos-II heat pipe system and its performance in test and orbit. Tech. Memo APL-TG-1049. Contract NOW-62-0604-C. Silver Springs, Md: Johns Hopkins Univ., Applied Physics Lab., 1969. AD 693 028.

4. HASKINS, W. L. Cryogenic heat pipe. Final report, Mar.–Dec. 1966, AFSC AFFDL-TR-66-228. Ohio: Air Force Flight Dynamics Lab. AFSC, Wright-Paterson Air Force Base, 1967. AD 657 025.

5. HIGGS, J. T., MUIR, N. A review of the literature currently available on heat pipes and their applications—Part I. BAC Mech. Office Information Sheet No. 493—Part I. Filton House, Bristol: British Aircraft Corporation Limited, Guided Weapons Division, Jan. 1969.

6. HIGGS, J. T., MUIR, N. A review of the literature available on heat pipes and their applications—Part II. B. A. C. Mech. Office Information Sheet No. 493—Part 2, 1969. Filton House, Bristol: British Aircraft Corporation Limited Guided Weapons Division, Jan. 1969.

1. JAKOB, M., FRITZ, W. *Forsch-Gebiete Ingenieurw.*, 1931, **2**, 434.

2. JOHNS HOPKINS UNIVERSITY, APPLIED PHYSICS LABORATORY. The Geos-2 heat pipe system and its performance in test and in orbit. NASA-CR-94585: JHU 52P-3-25. Contract NOW-62-0604-C. Silver Springs, Md: Johns Hopkins University, Applied Physics Lab., Apr. 1968.

3. JUDGE, J. F. R.C.A. test thermal energy pipe. *Missiles and Rockets*, Feb. 1966, 36–38.

1. KATZOFF, S. Notes on heat pipes and vapor chambers and their applications to thermal control of spacecraft. SC-M-66-623. Contract AT (29-1)-789. Proceedings of the Joint Atomic Energy Commission/Sandia Laboratories Heat Pipe Conference held at Albuquerque, New Mexico, 1 June 1966, pp. 69–89. Albuquerque, New Mexico: Sandia Corp., 1 Oct. 1966.

2. KATZOFF, S. Heat pipes and vapor chambers for thermal

control of spacecraft. AIAA Thermophysics Specialist Conference, New Orleans, Louisiana, 17–20 April 1967. Paper No. 67–310. New York, N.Y.: American Institute of Aeronautics and Astronautics, 1967.

3. KEMME, J. E. Heat pipe capability experiment. USAEC Ref. LA-3585-MS: Contract W-7405-eng-36. Los Alamos, New Mexico: Los Alamos Scientific Laboratory, Univ. of Calif., 1966.
4. KEMME, J. E. High performance heat pipes. USAEC Ref. LA-DC-9027. Contract W.-7405-eng-36. Los Alamos, New Mexico: Los Alamos Scientific Laboratory, Univ. of Calif., 1967.
5. KEMME, J. E. Heat pipe design considerations. ASAEC Ref. LA-4221-MS. Contract W-7405-eng-36. Los Alamos, New Mexico: Los Alamos Scientific Laboratory, Univ. of Calif., 1969.
6. KILMARTIN, H. E. The effect of wick geometry on the operation of a longitudinal heat pipe. M.Sc.Thesis. Monterey, Calif.: United States Naval Postgraduate School, June 1969. AD 692 442.
7. KINNEY, R. B., SPARROW E. M. Turbulent flow, heat transfer and mass transfer in a tube with surface suction. *J. Heat Transfer*, Feb. 1970, 117–125.
8. KNIGHT, B. K. JR., MCINTEER, B. B. Laminar incompressible flow in channels with porous walls. USAEC Ref. LA-DC-5309. Contract W-7405-eng-36. Los Alamos, New Mexico: Los Alamos Scientific Laboratory, Univ. of Calif., 1965.
9. KUNZ, H. R., LANGSTON, L. S., HILTON, B. H. *et al.* Vapor-chamber fin studies. Transport properties and boiling characteristics of wicks. NASA CR-812: Contract NAS 3-7622. West Hartford, Conn., Pratt and Whitney Aircraft Div. of United Aircraft Corp., 1967. Washington, D.C.: National Aeronautics and Space Administration, 1967.
10. KUNZ, H. R., WYDE, S. S., *et al.* Vapor chamber fin studies. Operating characteristics of fin models. NASA-CR-1139. Contract NAS 3-7622. East Hartford, Conn., Pratt and Whitney Aircraft Division of United Aircraft Corporation. Washington, D.C.: National Aeronautics and Space Admin., 1968.

1. LANDOLT-BÖRNSTEIN. Numerical values and functions. Volume 2, *Properties of matter in the various states of aggregation* (10 parts). Part 3 (1956), Melting point equilibria and

inter-face phenomena. Berlin: Springer-Verlag, 1956. Sixth edition.

2. LANGSTON, L., KUNZ, H. R. Vapour-chamber fin studies. First Quarterly Report, May 28–Aug. 28, 1965. NASA CR-54882, PWA-2698. Contract NAS 3-7622. East Hartford, Conn.: Pratt and Whitney Aircraft Division of United Aircraft Corporation. Washington, D.C.: National Aeronautics and Space Administration, 1965.
3. LANGSTON, L. S., KUNZ, H. R. Liquid transport properties of some heat pipe wicking materials, ASME Paper 69-HT-17. New York, N.Y.: American Society of Mechanical Engineers, 1969.
4. LAUB, J. H., MCGINNESS, H. D. Recirculation of a two-phase fluid by thermal and capillary pumping. JPL-TR-32-196. Contract NAS W-6. Pasedena, Calif.: California Institute of Technology, Jet Propulsion Laboratory, 1961.
5. LEE, J. D., WERNER, R. W. Concept for a gas buffered annular heat pipe fuel irradiation capsule. USAEC Ref. UCRL-50510. Contract W-7405-eng-48. Livermore, Calif.: Lawrence Radiation Laboratory, Univ. of Calif., 1968.
6. LEVY, E. K. Theoretical investigation of heat pipes operating at low vapor pressures. *Aviation and Space: Progress and Prospects:* Proceedings of the Annual Aviation and Space Conference, Beverley Hills, 16–19 June 1968, pp. 671–676. New York, N.Y.: American Society of Mechanical Engineers.
7. LUIKOV, A. V. *Heat and mass transfer in capillary porous bodies.* Translated from Russian by P. W. B. Harrison. Oxford, Pergamon Press, 1966, 524 pp.
8. LYMAN, F. A., HUANG, Y. S. Analysis of temperature distributions in heat pipe wicks. ASME Paper 69-HT-23. New York, N.Y.: American Society of Mechanical Engineers, 1969.
9. LYON, R. E., FOUST, A. S., KATZ, D. L. Boiling heat transfer with liquid metals. *Chem. Eng. Prog.*, Symp. Series, 1955, **51**(17), 41–47.
10. LYON, R. N. (editor). *Liquid–Metals handbook.* Pub. No NAVEXOS P-733 (Rev.). Sponsored by Committee of Basic Properties of Liquid Metals, Office of Naval Research, Dept. of Navy in collaboration with Atomic Energy Commission and Bureau of Ships, Dept of Navy, 1954. Washington D.C.: U.S. Govt. Printing Office.

1. MARTO, P. J., MOSTELLER, W. L. Effect of nucleate boiling

on the operation of low temperature heat pipes. ASME Paper 69-HT-24. New York, N.Y.: American Society of Mechanical Engineers, 1969.

2. MCADAMS, W. H. *Heat transmission*, 3rd edition. London: McGraw-Hill Book Co. Ltd., 1954.
3. MCKINNEY, B. G. An experimental and analytical study of water heat pipes for moderate temperature ranges. NASA-TM-X-53849. Alabama: Marshal Space Fight Center, June 1969; Washington: National Aeronautics and Space Agency, 1969.
4. MOSS R. A., KELLY, A. J. Neutron radiographic study of limiting planar heat pipe performance. *Int. J. Heat Mass Transfer*, 1970, **13,** 491–502.
5. MOSTELLER, W. L. The effect of nucleate boiling on heat pipe operation M.Sc. Thesis. Monterey, Calif.: U.S. Naval Postgraduate School, Apr. 1969.
6. MUSLIN Y. E. Superthermal conductivity. The heat tube. A thermal transformer. JPRS 47165. Washington, D.C.: Joint Pub. Res. Service, Dec. 1968.

1. NEU, H. The heat pipe, a new heat transfer system. *Eurospectra*, June 1970, **9**(2), 51–59.
2. NEW SCIENTIST. Heat pipe at work in a thermo-electric generator. *New Scientist*, 19 June 1969, **42**(654), 641.
3. NEW SCIENTIST. Heat pipes surge forward. *New Scientist*, 5 Mar. 1970, **45**(691), 461.

1. PALMER, W. B. The advance of a liquid front along a glass yarn. *Trans. Text. Inst.*, 1953, **44,** 391–400.
2. PARKER, G. H., HANSON, J. P. Heat pipe analysis. Adv. in Energy Conv. Eng. ASME, Intersociety Energy Conversion Eng. Conference, Miami Beach, Fla., 13–17 Aug. 1967, pp. 847–857. New York, N.Y.: American Society of Mechanical Engineers, 1967.
3. PEDERSON, E. S. Heat pipe thermionic reactor concept. *Nucl. Engng*, 1967, **12,** 112–114.
4. PEEK, R. L., MCLEAN, D. A. Capillary penetration of fibrous materials. *Ind. and Engng Chem.*, Anal. Ed., March 1934, **6,** 85–90.
5. PHILLIPS, E. C. Low temperature heat pipe program. NASA CR-66792. Contract NAS I-8000. Santa Monica, Calif.: McDonnel-Douglas Astronautics Co., June 1969. Washington: National Aeronautics and Space Agency 1969.

6. PHILLIPS, E. C., HINDERMAN, J. D. Determination of properties of capillary media useful in heat pipe design. ASME Paper 69-HT-18. New York, N.Y.: American Society of Mechanical Engineers, 1969.
7. PRODUCT ENGINEERING. Porous heat transfer surface boils liquids four times faster. *Product Engng*, 30 March 1970, **41,** 51.
8. PROVAN, T. F. Condensation of pure vapour in the presence of non-condensable gases: a survey to 1962. NEL Report No 114. East Kilbride, Glasgow: National Engineering Laboratory, Oct. 1963.

1. RANKEN, W. A., KEMME, J. E. Survey of Los Alamos and Euratom heat pipe investigations. USAEC Rept. LA-DC-7555: CONF-651049-3 Los Alamos, New Mexico: Los Alamos Scientific Laboratory, Univ. of Calif., 1965.
2. REISS, F., SCHRETZMANN, K. Pressure balance at maximum power density at the evaporation gained from heat pipe experiments. Inter. Conf. on Thermionic Electrical Power Generation, 2nd, Stresa, Italy, 27–31 May 1968, Proceedings. Conf. sponsored by European Nuclear Energy Agency. Luxembourg, EURATOM Centre for Information and Documentation (EUR No. 4210 f, e), 1969, pp. 507–513.
3. RIDEAL, E. K. On the flow of liquids under capillary pressure *Phil. Mag.*, 1922, **44,** 1152–59.
4. ROBERTS, J. J., CROKE, E. J., CARTER, P. P., *et al.*, Heat-pipe-cooled fast reactor space power supply. AEC Res 4 Dev. Rept. ANL-7422. Argonne, Ill.: Argonne National Laboratory, June 1968.

1. SAVAGE, C. J. Heat pipes and vapour chambers for satellite thermal balance. TR 69125.Farnborough, Hants: Royal Aircraft Establishment, June 1969.
2. SCHEIDEGGER, A. E. *The physics of flow through porous media: Solutions of Darcy's law.* Toronto Univ., University of Toronto Press, 1957.
3. SCHINS, H. E. J. Liquid metals for heat pipes, properties, plots and data sheets. EUR-3653 e. Ispra, Italy: Joint Nuclear Research Centre, Ispra, Italy, 1967.
4. SCHMIDT, E. Détermination de la perte de charge dans des milieux capillaires susceptibles d'etre utilises dans des caloducs. Note TT No. 265. Grenoble, France: Centre d'Etudes Nucleaires de Grenoble, 1967.
5. SCHMIDT, E., SEMERIA, R. Theoretical and experimental

determination of the limiting rate of heat transfer in sodium heat pipes (in French) Int. Conf. on Thermionic Electrical Power Generation, 2nd, Stresa, Italy, 27–31 May 1968. Proceedings. Conf. sponsored by ENEA. Luxembourg, EURATOM Centre for Information and Documentation (EUR No. 4210 e, f), 1969, pp. 515–527.

6. SCHWARTZ, J. Performance map of the water heat pipe and the phenomenon of non-condensable gas generation. ASME Paper 69-HT-15 New York, N.Y.: American Society of Mechanical Engineers, 1969.
7. SHLOSINGER, A. P. Heat pipe devices for space suit temperature control. Symp. Papers on Heat Pipes. Aviation and Space, ASME, New York, June 1968, pp. 644–648.
8. SHLOSINGER, A. P. Heat pipe devices for space suit control. NASA CR-1400. Contract NAS 2-3817. Calif.: TRW Systems Group, Redondo Beach, Oct. 1969. Washington, D.C.: National Aeronautics and Space Administration.
9. SEMERIA, R., SCHMIDT, E. Analytical and experimental study of sodium heat pipes (in French). Bulletin d'Informations Scientifiques et Techniques du Commissariat a l'Energie Atomique, Dec. 1968, No. 132, 31–48.
10. SILVER, R. S., WALLIS, G. B. A simple theory for longitudinal pressure drop in the presence of lateral condensation. *Proc. Instn. mech. Engrs*, 1965/1966, **180**(1), 36–42.
11. SILVERSTEIN, C. C. Preliminary evaluation of gas turbine regenerators employing heat pipes. USAAVLABS-TR-68-10. Contract DAAJO2-62-C-0053. Fort Eustis, Virginia: U.S. Army Aviation Materials Labs., April 1968.
12. SILVERSTEIN, C. C. A study of heat pipe applications in nuclear aircraft propulsion systems—Final report. NASA-CR-72610: Contract NAS 3-11841. Cleveland, Ohio: Lewis Research Center; Washington, D.C.: National Aeronautics and Space Administration, Dec. 1969.
13. STENGER, F. J. Experimental feasibility study of water-filled capillary pumped heat transfer loops. NASA-TM-X-1310. Washington, D.C.: National Aeronautics and Space Administration, Nov. 1966.
14. STRECKERT, J. H. *et al.* Development of a versatile system for detailed studies of the performance of heat pipes. NASA-CR-100725. ME-TR-64. Urbana, Ill.: University of Illinois, Department of Mechanical and Industrial Engineering, Dec. 1968. Washington, D.C.: National Aeronautics and Space Administration.

1. TIEN, C. L. Two-component heat pipes. AIAA Paper

69–631. American Institute of Aeronautics and Astronautics 4th Thermophysics Conference, San Francisco, Calif., 16–18 June 1969.

2. TURNER, R. C. The constant temperature heat pipe—a unique device for the thermal control of spacecraft components. AIAA Paper 69–632. American Institute of Aeronautics, and Astronautics 4th Thermophysics Conference, San Francisco, Calif., 16–18 June 1969.
3. TURNER, R. C., HARBAUGH, W. E. Design of a 50,000 W heat pipe space radiator. *ASME-Aviation and Space: Progress and Prospects*. Annual Aviation and Space Conference, Beverley Hills, Calif., 16–19 June 1968, 639–643.

1. VAN ANDEL, D. Heat pipe design theory. Int. Conf. on Thermionic Electrical Power Generation, 2nd, Stresa, Italy, 27–31 May 1968. Proceedings. Conference sponsored by European Nuclear Energy Agency. Luxembourg, EURATOM Centre for Information and Documentation (EUR No 4210 e, f), 1969, pp. 529–542.

1. WAGEMAN, W. E., GUEVARA, F. A. Fluid flow through a porous channel. *Physics of Fluids*, 1960, **3,** 878–881.
2. WALLIS, G. B. Pressure gradients for air flowing along porous tubes with uniform extraction at the walls. *Proc. Instn. mech. Engrs.* 1965/1966, **108**(1), 27–35.
3. WALLIS, G. B. Discussion reference K7.
4. WASHBURN. E. W. The dynamics of capillary flow *Phys. Rev.*, 1921, **17,** 374–375.
5. WATTS, J. L. High temperature properties of liquid metals. USAEC Ref UCID-15176. Space Power Note No 221. Contract W-7405-eng-48. Livermore, Calif.: Lawrence Radiation Laboratory, Univ. of Calif., July 1967.
6. WATTS, J. L. A heat pipe optimization code, LAM 2. UCID-15462. Livermore, Calif.: Lawrence Radiation Laboratory, Univ. of Calif., April 1969.
7. WERNER, R. W., CARLSON, G. A. Heat pipe radiator for a 50 MW space power plant. USAEC UCRL-71004. Contract W-7405-eng-48. Livermore, Calif.: Lawrence Radiation Laboratory, Univ. of Calif., 1968.
8. WHITING, G., FELDMAN, K. T. JR. Applicability of heat pipes to energy conversion systems. SC-ARPIC-1012. Livermore, Calif.: Aerospace Radioisotope Power Information Center, Sandia Labs., June 1969.
9. WYATT, T. A controllable heat pipe experiment for the

5 E-4 satellite. JHU-APL-SDO 1134. Contract NOW-62-0604. Maryland: The Johns Hopkins University Applied Physics Lab., Mar. 1965. AD 695 433.

1. YUAN S. W., FINKELSTEIN, A. D. Laminar pipe flow with injection and suction through a porous wall. *Trans ASME*, 1956, **78,** 719–724.

Appendix 1: Conversion of Units

Pressure

$1\ \text{lb/in}^2 = 0{\cdot}06895$ bar
1 torr $= 0{\cdot}001333$ bar
$1\ \text{N/m}^2 = 10^{-5}$ bar

Quantity of Heat

1 Btu = 1055 J

Rate of Heat Transfer

1 Btu/h = 0·2931 W

Heat Flux

$1\ \text{Btu/h.ft}^2 = 3{\cdot}155\ \text{W/m}^2$

Heat Transfer Coefficient

$1\ \text{Btu/h.ft}^2\text{.deg F} = 5{\cdot}678\ \text{W/m}^2\text{.K}$

Density

$1\ \text{lb/ft}^3 = 16{\cdot}02\ \text{kg/m}^3$

Specific Latent Heat

1 Btu/lb = 2·326 kJ/kg

Surface Tension

1 lb/ft = 14·594 N/m.

Appendix 2: Manufacturers of Heat Pipes

Dynatherm Corporation,
One Industry Lane,
Cockeysville, Md. 21030.

Energy Conversion Systems, Inc.,
3821 Commercial, N.E.,
Albuquerque,
New Mexico 87107.

Heat Pipe Corporation of America
Division of Woods-Palmer Corporation
141 Park Place,
Watching,
New Jersey 67060.

McDonnel Douglas Astronautics Co.,
Donald W. Douglas Laboratories,
2955 George Washington Way,
Richland,
Washington 99352.

Noren Products, 846 Blandford Blvd.,
Redwood City,
California 94062.

RCA, Heat Transfer Devices Marketing,
Lancaster,
Pa. 17604.

RCA Limited,
Sunbury-on-Thames,
Middlesex.

Thermo Electron Corporation,
85 First Avenue,
Waltham,
Massachusetts 02154.

INDEX

M & B MONOGRAPHS

MECHANICAL ENGINEERING

ME/1	**Tribology** E D Hondros	0.263.51588.5
ME/2	**Electron Beam Welding** M J Fletcher	0.263.51707.1
ME/3	**Vacuum Brazing** M J Fletcher	0.263.51708.x
ME/4	**Stress Corrosion Failure** P Greenfield	0.263.51725.x
ME/5	**Glow Discharge Material Processing** R A Dugdale	0.263.51787.x
ME/6	**Diamond Grinding** J Burls	0.263.51791.8

CHEMICAL ENGINEERING

CE/1	**Microbiological Corrosion** G H Booth	0.263.515842
CE/2	**Solvent Treatment of Coal** W S Wise	0.263.51706.3
CE/3	**The Fluidised Combustion of Coal** D G Skinner	0.263.51714.4
CE/4	**Carbonisation of Coal** J Gibson D H Gregory	0.263.51774.8
CE/5	**The Principles of Gas Extraction** P Paul W S Wise	0.263.51802.7
CE/6	**Metal-Air Batteries** D P Gregory	0.263.51838.8

ELECTRICAL ENGINEERING

EE/1	**Threshold Logic** S L Hurst	0.263.51587.7
EE/2	**The Magnetron Oscillator** E Kettlewell	0.263.51641.5
EE/3	**Material for the Gunn Effect** J W Orton	0.263.51773.x
EE/4	**Microcircuit Learning Computers** I Aleksander	0.263.51724.1
EE/5	**D C Conduction in Thin Films** J G Simmons	0.263.51786.1
EE/6	**Reading Machines** J A Weaver	0.263.51839.6

M & B TECHNICAL LIBRARY

TL/ME/1 Hydrostatic Extrusion 0.263.51709.8
J M Alexander
B Lengyel

TL/ME/2 The Heat Pipe 0.263.51711.x
D Chisholm

TL/CE/1 Boron 0.263.51775.6
A G Massey
J Kane

TL/EE/1 Linear Electric Motors 0.263.51586.9
E R Laithwaite

TL/EE/2 An Introduction to the Josephson Effects 0.263.51705.5
B W Petley

FORTHCOMING TITLES

Engineering Applications of Beryllium
P Greenfield

Whiskers C C Evans

Creep of Metals P Greenfield

Cavitation I S Pearsall

Fuel Cells D P Gregory

Speech Synthesis J N Holmes

Glow Discharge Display G F Weston